GW00504244
S G Woods

This is number 109

of a limited edition of 250 copies.

C. A. STRAIN,
Publishing Director.

MOTAT
Museum of Transport and Technology
of New Zealand (Inc.)

To Sir Ralph Cochrane

With the compliments & regard from all at Motat.

H. Richardson
Director 7-10-77

Front cover:

Motat's 1912 Brush veteran car outside Willow Cottage. Len Cobb

Back cover:
Photographed from under the wing of the RNZAF Sunderland, looking towards the Teal (now Air New Zealand) Solent ZK-AMO.

Published by Paul Hamlyn Limited,
Levien Building, St Paul Street,
Auckland, New Zealand.

First published 1976

Typeset in New Zealand by Jacobson Typesetters Ltd
Printed by Toppan Printing Co. Ltd.
38 Lui Fang Road, Jurong Town, Singapore 22

ISBN 0 600 07423 4

Designer: Leonard Cobb

MOTAT

Museum of Transport and Technology
of New Zealand (Inc.)

compiled by
John Cresswell

HG PAUL HAMLYN
AUCKLAND SYDNEY LONDON NEW YORK TORONTO

FOREWORD

by His Worship the
Mayor of Auckland,
Sir Dove-Myer Robinson

In a few short years Motat, the Museum of Transport and Technology of New Zealand, has become one of the major tourist attractions of Auckland and indeed of this country. Even by world standards the organisation is unique in respect of the live nature of its displays and the involvement of its large number of voluntary working members.

The success of the venture is a tribute to a few far-sighted people from all walks of life, who persevered and battled through the formative years. Their rewards are manifest in an undertaking of which today it can be truly said that 'history comes to life'.

Although largely self supporting, the museum has earned the goodwill and co-operation of my Council, the Government and the business community of this city and the country as a whole.

Motat's greatest accolade is the ever increasing involvement of the general public, who visit it again and again.

CONTENTS

Things are moving at Motat during a Live Weekend.

MOTAT — MUSEUM OF TRANSPORT AND TECHNOLOGY OF NEW ZEALAND (INC.)

INTRODUCTION

If it works or can be made to work, save it. This is the criterion for the mechanical exhibits displayed at the Museum of Transport and Technology of New Zealand, commonly known as Motat. It is observed today with the same spirit as it was by the dedicated band of collectors, technical enthusiasts, and amateur historians who formed the nucleus of the museum in its formative stages. Regular Live Weekends, when tram, buses, aeroplane engines, and other vehicles are put through their paces, are thus possible.

The story of the development of the museum began on June 5, 1960, when the Old-time Transport Preservation League of Matakohe, the Royal Aeronautical Society (NZ Division), and the Historic Auckland Society convened a meeting which was presided over by the Mayor of Auckland, (later Sir Dove-Myer Robinson). The aim of the meeting was to bring together various historical collections. The result was a museum which preserves historical forms of transport and pioneer artifacts in order to record the advances of technology.

The next 10 years were vital to the success of the museum. Old things had not yet acquired intrinsic 'antique' value, and much of what museum members saved would otherwise have been lost forever. Much had already been sacrificed to war-time scrap drives. From barns and attics and wrecker's yards, from phased-out tramways and steam railways, from airline, fire brigades and farms and even from buried dumps of war-time aircraft, the museum's supporters brought in thousands of items, large and small. And once the project began to become news the public responded with a further flow of relics, heirlooms, and even whole private collections.

The three-hectare site at Western Springs, Auckland, was a unanimous choice for the museum. But storage was a problem, and many wooden buildings, mostly ex-Army, were donated for this purpose. Even after the official opening of Motat in October, 1964, material was stored in old, hastily renovated buildings, under tarpaulins and shelters, small items inside large vehicles. Protection was the key note, display at that stage subordinate to preservation and restoration. The organisation, identification, description, and preliminary restoration of all this material was a huge task. There was little room for planned professionalism, and little professional expertise available in what was, for New Zealand, a unique undertaking.

Working like beavers in this accumulation of treasure were scores of enthusiasts with varied interests and skills, from retired technicians and tradesmen to boys, often their sons, learning by doing. Groups formed to specialise in sections such as Trams, Rail, Agriculture, Aviation and Radio. Workshop facilities were set up to clean and repair, and for some years these workshops had to double as display areas. Leaders emerged who could produce miracles in the way of restoration with negligible facilities or finance — even to laying tram tracks and reviving trams to operate at acceptable safety standards, rebuilding aircraft and locomotives, setting up working exhibits of primitive radio equipment, and demonstrating old farm machinery. The characteristic ingenuity and versatility of the New Zealander extended to the use of some items acquired as exhibits — old tractors and other obsolete equipment — in the work of progressively developing the rugged site.

While the first Executive Committee and Director pored over information and advice from the world's great sophisticated museums such as Britain's Science Museum and America's Smithsonian Institution, seeking guidance in planning and organisation, their domain bore a closer resemblance to Steptoe and Son. They strove to obtain finance, Government and public recognition, adequate buildings, and to achieve reasonable standards of presentation and useful educational services. Most museums start with Government or local Government backing, financial and otherwise, professional staff, and already established collections. Motat had none of this. It had to scrounge nearly everything required at a time when most of the good-natured donors had no real confidence in the viability of the project and just felt that so much enthusiasm deserved some support.

Motat is now permanently established, with some fine buildings and displays, wide public support, even international status and experienced — if unusually small — staff. Volunteer groups continue to maintain, improve, and operate some of the exhibit sections and many of the same faces can still be seen that were struggling a decade ago with the raw material of this impressive institution. Boys have acquired skills and become men in the unpaid service of the museum.

There have been problems, predicted by other museum administrators, but enthusiasm is a jewel without price and the results are evident. Almost a generation has now felt the impact of this dynamic revival of technological traditions. Generations to come will enjoy the experience and know little of the pioneering efforts that made it possible.

Contributors to this book:

C.J.W. Barton
A.T. Blampied
R.S.L. Downey
B.W.C. Forster
M. Frazer
C. Harding
E. Harding
J.H. Hogan
J.C. Lewis
Mrs. W. Macdonald
J.H. Malcolm
F.J. Neil
Mrs I.E. Pirrit
D. Reddick
R.J. Richardson
W.V. Robinson
B.R. Skudder
L. Stenersen
M.D. Sterling
I.W. Stewart
P.J. Tibbutt
E.J. Wansbone
S. Waterman
Mrs R. E. Wright
F. Wright

In addition to the above major contributors, many museum members and staff have made their own contributions to this book. Their efforts are recognised by this general acknowledgement. Special thanks are accorded to Christine Harris for her tactful editorial supervision of a much fragmented manuscript and to Nanette Cresswell for her proofreading.

GUNSON'S FERTILE
SEEDS
HARDING STABLES
BIG TREE
MOTOR OIL

1

Graham C. Stewart

1. 1891 Baldwin ex Sydney Steam tram No. 100

ROAD TRANSPORT

The development of the motor car from a horseless carriage to the sophisticated modern automobile and from the early hard-tyred motor lorries to the efficient monster haulers of today is a fascinating story of the ingenuity of man.

Motat's collection of 61 vehicles (not including fire engines) dating from 1903 onwards includes some rare veteran and vintage models. Twenty-seven of these are completely restored and in good working order, others are under restoration, and the rest lie in storage awaiting their turn. It is seldom possible to restore more than two vehicles per year, and some have been and will continue to remain in storage for several years awaiting attention. However, with the prices of veteran and vintage cars ever increasing, the policy of collecting and storing for future restoration has proved a wise move.

Museum restoration involves virtually a complete rebuilding. It is a long and costly job involving countless man hours of painstaking skilled labour. Considerable ingenuity goes into it, for with parts being practically unobtainable most replacements are made at the museum. The skill of an expert fitter and turner on the permanent staff is always in great demand.

Once restored, the vehicles require a great deal of expert attention to maintain them in first class running order. A constant battle must be waged against corrosion and other ravages of time. The adequate housing or restored vehicles is a constant problem, and they are frequently rotated from storage to exhibition.

As Motat's restored collection grows, rallying cars in veteran and vintage events is popular amongst members, who take great pride in the fruits of their labour. Street parades in Auckland City have included vehicles from the museum, and some members like to be married in the Pioneer Village Church and use vintage bridal cars. There is also an increasing demand for cars to be used in television presentations — a valuable source of income to offset the high cost of restoration.

There are two fine veteran truck chassis awaiting restoration, one a 1914 Commer Car and the other a 1915 F.W.D. London Truck,

Opposite page:

Horses and carts outside the museum's stables. These stables were built in 1910 and were a landmark at St. Heliers Bay before being transported to Motat and re-erected in their original context.

Cobb

1

Len Cobb

the latter being an American World War I 'lend-lease' vehicle. Both of these vehicles are runners, and on occasions members crank the monsters' engines into life with the aid of copious applications of cold-start spray. Some have suffered sprained wrists and even a broken arm for their trouble.

Another impressive item is a large black Daimler Hearse. Before conversion to a hearse it was a state limousine with a transparent roof, and it was used by Queen Elizabeth II during her first tour of New Zealand in 1953.

A long-term restoration job is the 1923 White Bus. This was used for many years on the North Shore of Auckland, transporting commuters to and from the ferry boats. It was retired to Rangitoto Island where it was the only bus in service. In 1968 it was given to the museum and brought by barge to the city to be driven to its final home.

Other commercial vehicles include a rare 1942 Reliant three-wheeled van, possibly the only one in New Zealand, and a 1948 Austin Beer Tanker which has the honour of being the first of its kind in this country and possibly in the world.

1. Girlie contemplates a busy day ahead amongst the thousands of children.

2. Remember the days of the horse float? Horses hooves, steel wheels, and the jangle of milk cans.

Motat's horse-drawn vehicle section is as much alive as the motor transport. Four horses have been acquired and are regularly employed hauling a variety of equipages giving both children and adults a sample of travel in an age that has gone forever. The sight of the museum horses being ridden along Great North Road to the main gates on weekends, each with a girl volunteer astride, is well known to Aucklanders in the Western Suburbs.

Motat possesses an interesting sample of vehicles representative largely, but not entirely, of the heyday of horse-drawn transport in the country. Horse-drawn transport had a long history with new and improved vehicles appearing over the years. It should be remembered that the evolution of the horse-drawn vehicle took place in close association with that of the roads. The process was a two-way one. As vehicles became more sophisticated they required better roads on which to run, and as roads improved more and better vehicles were produced to match the improvements.

As is the case with the internal combustion engine today, the horse in New Zealand was put to a variety of uses. It was a means of private transport of both families and their goods, particularly, but by no means exclusively, in country districts, and a means of public transport of both passengers and freight. The horse was also used for specialised haulage tasks for farms, forests, shopkeepers, contractors, industrial and commercial concerns, fire brigades, funeral directors, and for various tasks associated with the construction of the roads themselves. Motat has examples of vehicles and appliances associated with all of these.

In the heyday of the era of horse transport in New Zealand there were several equivalents of the private car of today. First there was the gig — a light, two-wheeled, single-seated vehicle for two people (or three with a squeeze). It usually had room under and perhaps behind the seat for parcels and small luggage, and it was drawn by a single horse between two shafts.

The trap (or cart, or dog cart) was a somewhat larger and more solidly built vehicle, sometimes with room for two more passengers, sitting back to back with those in front, and for more luggage.

2

Motat collection

L. Cobb

1

1. Girlie, hitched to a four-wheeled buggy, patiently awaits her passengers outside the museum's general store.

Motat collection

1

Motat collection

2

In the governess cart, passengers sat facing each other along each side (four adults or somewhat more children). Its name comes from the original intention that it was for the use of children with an adult driver.

In addition to these two-wheeled vehicles, which were by far the most numerous, there were the four-wheelers. These were the buggies, including the superior phaetons, landaus, and broughams as well as more run-of-the-mill types. There were both one- and two-horse versions. Motat's collection of private passenger vehicles consists of twelve two-wheeled vehicles and six four-wheeled.

For private and public goods transport, as in the case of private passenger vehicles, once again there were both two- and four-wheeled types. The former are represented by the dray, the latter by the wagon. Drays in New Zealand were principally of two kinds. One was a smaller tip dray commonly used to transport bulk loads of shingle, sand, firewood, and farm crops such as mangolds. These played a major role in farm operations and public works. They were also used as non-tip drays for small loads of bagged potatoes, cream cans, and so on.

The other dray, the spring dray, was larger and normally not of the tip variety. It also played an important role on the farm — during haymaking and harvesting, for example — and was extensively used by general carriers, coal and firewood merchants, and others.

1. The beginnings of the collection — horse-drawn buggies and coaches stored at Matakohe.
2. Very early days of Motat — wagons arrive from all parts of the country.

The four-wheeled freight vehicle was the wagon. Its functions were similar to those of the larger drays, but it could handle much bigger loads. Drays were usually hauled by a single horse, although on bad roads and muddy farms a second horse was sometimes attached, either alongside the first or in tandem. Wagons of the larger kind were usually hauled by two horses or, in the case of the very big ones, such as furniture vans, by a team of four or more — a stirring sight. It should be added that large powerful horses — draught and half-draught — were used with drays and wagons. Motat's collection consists of five drays and seven wagons.

The 'glamour section' of transport provided by the horse was the public passenger transport. For this reason much has been written about it, and it has been extensively portrayed in pictures, photographs, and films. The long-distance stage coach is familiar to all, with passengers inside and outside and its team of sure-footed horses. Less familiar to today's generations are the short-distance, less picturesque coaches, the urban and suburban ones (and horse-drawn trams), and the 'drags' that met trains at railheads and other stations. And we must not forget the predecessor of the taxi, the hansom cab.

Because of the gaps in the railway network during the early part of this century (in the Main Trunk in the North Island, for example, and in the Midland Line in the South Island), and because motor transport was somewhat slow in its growth in New Zealand, horse-drawn passenger transport persisted longer than one might have imagined. The mail coach was still crossing Arthur's Pass in the 1920s, and hansom cabs were still to be seen in some cities at this time.

Motat has restored a four-horse-team Royal Mail Coach, a hansom cab, and the nine-seater Waiwera Buggy.

Other horse-drawn 'vehicles' were used on the farms in the shape of agricultural implements. And butchers and bakers had strange box-like, two-wheeled carts in which their wares were taken to customers, free from the dust of unpaved roads. Milkmen used two-wheeled floats, open to the elements, containing their cans. In the large cities, semi-enclosed four-wheelers were also sometimes used, persisting in Wellington until well into this century and in Auckland to the early forties.

At Motat there are four baker's carts, two milk floats, and a hearse, all horse-drawn, and, to include the early ally of the horse on farm and public works and in the forest, one bullock dray. The museum keeps several horses for use with these vehicles.

Motat collection

Motat collection

Opposite page:

Girlie and Ben bringing joy to passengers who had almost forgotten what it is like to ride in a horse and cart.

1. Motat's team, Ben and Girlie, hitched up to the historic Waiwera nine-seater buggy.
2. Transport of yesteryear. Members of Motat in period costume in the 1920-era shopping street. In the background, the cheapest butcher shop in town displays genuine 1913 meat prices.

1
2

Motat collection

BRUSH

Veteran & Vintage Cars

The Auckland Veteran and Vintage Car Club was invited to help with the development of Motat at a very early stage in its formation. The club was involved in many activities not directly related to vintage cars, and its first major restoration was the 1905 Hornsby Hot Bulb engine which is one of the noisiest attractions of Live Weekends. Members of the club also restored a McCormack Deering tractor and a c.1890 Union petrol engine, which was once used to power a Northland sawmill.

The next and largest restoration job was the Renault charabanc. This monster vehicle was discovered beneath some gum trees on a farm property at Karaka. A search for missing pieces revealed the dash-mounted petrol tank under a dead sheep, but other important parts were not located. It was purchased for a shilling.

Inspired by a picture of an early fifteen-seater Renault that appeared in a New York magazine, members christened the wreck *Angelique*. It had been a timber lorry among other things, but they decided to rebuild it as a bus. Drawings were based on the original French designs, and the wood frame construction was approved shortly before the law for body materials was changed in 1967.

Many adaptations of modern components had to be made to restore the vehicle. Differential bearings were imported from Sweden, and modern brake linings were fitted instead of the original cast-iron pads. The engine was started for the first time in 1966. The body was built up and professional upholstery installed — the only paid labour on the whole job. Finally *Angelique* was ready for the road and made her debut in the Auckland City Centennial parade in 1971.

The restoration is dedicated as a memorial to a notable veteran car enthusiast, the late Horace Robinson, who is a brother of Sir Dove-Myer Robinson, Mayor of Auckland and patron of Motat.

In December 1972 tragedy struck the museum in the form of a fire which practically gutted the shed which housed *Angelique*. Luckily the blaze occurred during the day and was noticed early. In spite of the long delay caused by running hose over 30 metres from the nearest hydrant and by the large crowd at the museum, the fire brigade made a good save. Six years of patient restoration was lost in as many minutes — *Angelique* was a sorry sight with upholstery and woodwork charred and paintwork and tyres burnt.

1. Mr White of Skeates & White with the penny farthing bicycle donated to the Old-time Transport Preservation League.

Opposite page:

Motat's 1912 Brush veteran car outside Willow Cottage — Model E Everyman's Car, 7.5 kW single cylinder, 50 km/h. Originally priced at $450. Built by the Brush Runabout Company, Division of the United States Motor Company, Detroit, Michigan.

It took over a year to restore the vehicle to its former glory, with one member working on her for several months. Now fully restored again, it is a recognised fact that the charabanc has great trouble passing an open pub. Many a publican has had a dozen thirsty crew members suddenly descend upon him while his other bar patrons go outside to admire the resting, shiny blue *Angelique*.

Among the cars owned by Motat, much of the limelight is stolen by the 1910 International High Wheeler Auto Buggy Model D,975. Six thousand vehicles of this type were made, and this particular car was purchased by a Mrs Goodin of Carterton in September, 1910, through the local agent, Mr R. B. Forsyth. It cost her £260. The car was later sold and taken to Te Kauwhata. It eventually arrived in Auckland in the late 1930s.

n Cobb

1

Auckland Star

1. The burnt-out remains of *Angelique*, the 1911 Renault charabanc. Six years of patience lost in as many minutes.

1

Motat collection

1. Children love to ride in *Angelique,* the 1911 Renault charabanc, similar to the type of vehicle used as troop transports in the Fields of Flanders during World War I.

The International has a twin-opposed engine with 12.5 cm bore and 12.5 stroke which develops 15 kW at 1800 r.p.m. The crankshaft runs on phosphor-bronze brushes and the drive is by a chain from gearbox to countershaft to both rear wheels. The front wheels are 102 cm high and the rear 112 cm with tyres 4.5 x 3.8 cm. It has a top speed of 16 km/h and does 3.5 km to the litre. There is seating for four.

Visitors to Motat invariably gather around the fascinating 1908 Sizaire et Naudin veteran racing car, which contrasts with the Cooper Climax Formula I racing car formerly owned by the famous New Zealand racing driver, the late Bruce McLaren. The old racer has a number of interesting features including independent front suspension and three-speed differential. It also has a wooden chassis. It was driven to win a number of races, the first being on the famous Sicilian Targo Floria circuit.

A recent acquisition to the veteran and vintage collection is the Black & White Taxicab, which is a 1928 Studebaker sedan car restored in the colours of a bygone Auckland taxi company with the assistance of the Auckland Co-Operative Taxi society. The Society incorporates all the taxi companies which previously operated in Auckland: Atta, Auto, Black & White, Checker, Chess, Diamond, Red Bond, and Red Top.

The Brush Runabout Model E Everyman's Car was built 1911-12 by the Brush Runabout Company of Detroit and was priced at $450. Only recently brought to New Zealand, it is still in fine working order. It has a 7.5-kW, 10-cm, single-cylinder motor and can reach speeds of nearly 50 km/h.

Two very popular cars of the 1920s are represented by a 1926 Ford Model T Touring Car *New Beauty* and the 1929 Essex, a typical medium-priced sedan car of the period. Its six-cylinder engine features a cork-plated fluid clutch. A similar vehicle, the 1926 Nash sedan, is displayed unrestored in the Transport Pavilion as an indication of the condition in which most vehicles are received. The cost of restoring the vehicle would be in excess of $3000 and approximately 2000 man-hours.

1. Cooper Climax Formula I racing car, the car in which the late Bruce McLaren won the Monte Carlo Grand Prix.

2. 1908 Sizaire et Naudin veteran racing car, top speed 80 km/h.

1

Motat collection

2

Motat collection

Motat collection

Barry McKay

2

Motat collection

3

1. Prized exhibit — 1908 International Auto Buggy.

2. Motat's collection of vintage vehicles slowly builds up. A scene soon after the opening of the museum to the public.

3. Shades of 1921. Beer deliveries by Ford Model T brewery truck.

1

2

Motat collection

1. Motat's veteran cars are all in operational condition and are taken for frequent outings.

2. 1912 Duo Cycle Car (England). Two-cylinder air-cooled JAP engine, gearing achieved by expanding pulleys and movable back end. The only known survivor of its type in the world.

The oldest car in Motat's display is a 1903 Oldsmobile. A 1913 Wolseley Landau 16/80 Gentleman's Town Carriage is an exceptionally fine exhibit from the days of gracious motoring.

At the other end of the scale and a little later than vintage is the tiny Messerschmitt Bubble Car with a Sachs 191 c.c. engine. These little vehicles enjoyed a certain popularity in the immediate post World War II period.

In the odd and unusual category there is the 1912 Duo Cycle Car from England. This has a two-cylinder JAP air-cooled engine. Gearing is achieved by expanding pulleys and a movable back end. This is the only known survivor of its type in the world.

The famous Baby Austin is represented by a 1926 Austin Roadster, displayed in the Transport Pavilion on a one-way bridge depicting the advances made in road-making and bridge-building throughout New Zealand in a relatively short period. Also in the Pavilion is a display of early motorcycles, including a 1928 Harley Davidson and sidecar, a 1911 Douglas, and a Villiers motorised bathchair of about 1930. There is also a cutaway demonstration unit featuring an N-Zeta motor scooter, the first motor scooter manufactured entirely in New Zealand.

Military vehicles are represented by a 1916 Rover Sunbeam Ambulance, a 1942 Ford V8 Desert Ambulance, and a 1942 Willys Jeep, the standard United States all-purpose vehicle of World War II.

The Rover was used in France during World War I. It was shipped to New Zealand where it was stationed at Greymouth, for a long time the only motorised ambulance on the West Coast. It was discovered by Motat at Kerikeri, Northland, with a tamarillo tree growing through the bonnet. Restored to war service condition, it again bears the ambulance marking of the Royal Army Service Corps in France.

The Ford Ambulance was used during World War II by the Second New Zealand Expeditionary Force in the Western Desert. It was restored for Motat by the New Zealand Army Workshops.

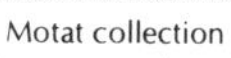

Motat collection

1
2

3

Motat collection

1. From the age of elegance — 1913 Wolseley Landau, known as The Gentleman's Town Carriage, glass-partitioned and complete with speaking tubes.
2. Motat's beautifully restored 1905 Model T touring car, *New Beauty*.
3. Remember the days when . . . Senior citizens share their memories of other days.

Motat collection

Motat collection

1

1. The late Miss R. Kelly, nursing sister during World War I, stands beside a 1915 Rover Sunbeam ambulance which saw service in France.

2. 1928 Studebaker sedan restored as a Black & White taxicab of the period.

3. The McLaren M6BGT, the late Bruce McLaren's first attempt at designing and building a sports saloon car. Finished late in 1969, it was used on the road in and around London for 9 months or so until Bruce's tragic accidental death.

4. An early Model T tiptruck, a development from the days of shovelling on, and shovelling off.

Opposite page:

Patron of the museum, Sir Dove-Myer Robinson, filling in as clippie on the London double-deck bus.

Motat collection

4

2

Motat collection

3

Motat collection.

10c

1

Motat collection

2. Motat collection

Buses

The introduction of the London double-decker buses at Motat was due to the generosity of Waikato Breweries and its chairman of directors, Mr Harold Innes. The brewery sponsored and imported the first bus in 1971 to coincide with the local launching of the internationally known beer, Bass. This vehicle became known as the Bass Bus. It was a sensation with museum visitors and very soon proved its worth as a fund-raiser. In fact the bus became so popular that another vehicle became necessary, and the brewery again came forward with help, sharing the cost of purchase and shipping. The second bus, which is now called the Worthington Bus after the advertisements it carries for another famous beer, went into service at Motat in December, 1972.

The buses are operated by a poole of members who are qualified omnibus drivers, and others assist as conductors and conductresses. Buses run on a regular schedule during weekends and holidays and on demand during the week. The museum is licensed to operate a passenger service over a route which passes the Meola Road Aviation Park extension of the museum. The buses are fitted with a public address system for describing the aeroplanes to passengers as they pass by. In addition, the buses operate on charter trips outside the museum, both for the public and for members taking a break from the chores of the museum. The charter trips include school and church fairs and fetes, charity fund-raising promotions, old folks' outings, and so on.

During the two years after the arrival of the first bus in 1971, with the second bus in its first year, the two vehicles transported 410,126 passengers (mostly on the top deck) 61 785 kilometres. This included 58 special brewery promotions and 74 charity trips. They also worked for 312 promotions outside the museum and carried Father Christmas to 18 functions. All this earned $23,000 net toward restoration and improvements at Motat. A very widely recognised feature of the museum, they were shown on television 23 times and in newspaper and magazine photographs on 57 occasions during this period.

Built in 1948 the buses are R.L.H. models, about 35 cm lower than the average London bus. They were built for service on routes where there were low overhead bridges and wires to negotiate. They are A.E.C. vehicles with 72-kW diesel engines and pre-selective gearboxes. Each seats 53 and retains its original designation number, RLH 45 and RLH 50.

1

Motat collection

2

Motat collection

3

Motat collection

Opposite page:

1. Motat's Christmas present from Bass International Ltd of London being unloaded at the Auckland Wharves, November, 1971.
2. Motat's first London double-deck bus arrives at the museum, November 5, 1971.

1. 1923 White motor bus on its final journey into retirement at Motat after 45 years of hard labour.
2. Repainted in its original configuration as North Shore Transport Company No. 4, the White bus in the Auckland Centennial Parade.
3. Motat's second London bus on charter, October, 1973. No shortage of passengers upstairs.

Bass
149
Dalston Shoreditch
Bank Southwark Bdg
Waterloo Station
Lambeth Bridge
Bass
Bass
the great beer of England!
RLH 45
LONDON TRANSPORT
GJ 3835
BUS STOP

Transport Pavilion

The Transport Pavilion was made possible by the generosity of the Auckland Savings Bank which presented the building to Motat and the City of Auckland to mark the centennial of the Auckland City Council in 1971. The contemporary design of the architects, Adams & Dodd, has resulted in a pleasant combination of brick, steel, and glass, which has created the desired effect of space and light, breaking away from the traditional museum design. The $130,000 pavilion, the first permanent building to be erected at the museum, was opened in 1972.

The transport exhibition has been planned to illustrate the contrast between veteran, vintage, and modern forms of transportation on land and sea and in the air. The entrance to the pavilion features a credit board commemorating the gift of the building and the origins of the museum. Opposite is a collection of five original paintings by the eminent New Zealand artist-historian Mr W. W. Stewart. They depict the five basic means of transport.

The painting exhibits include a K900 locomotive, a Wellington tramcar No. 252, a Solent Flying Boat, *RMS Aranui,* (which stands in the museum's Aviation Park, Meola Rd), the ferryboat *Makora,* and a 1912 Austin motor car owned by Mr Wood of Mt Eden. Exhibits in the Transport Pavilion change, subject to special displays commemorating anniversaries in the transport history of New Zealand, to the normal rotation of exhibit material from storage areas, and to the loan of exhibits to institutions elsewhere in the country. Change is also dictated by the restoration programme and the display of newly acquired veteran and vintage vehicles. Each current display at the pavilion is described in the Transport Pavilion Catalogue, which is mailed on request or can be obtained from the souvenir shop at the museum.

The predominate theme of the exhibition will always be the contrast between vintage and more modern technological developments in transportation. The display incorporates scenes of New Zealand transport history and commemorates the achievements of famous New Zealanders.

A diorama depicts the historic trek to the South Pole by Sir Edmund Hillary and Peter Mulgrew, perhaps one of the greatest transport endurance feats of modern times. The expedition departed from Scott Base 14 October, 1957, and arrived at the South Pole 4 January, 1958, after covering a distance of over 2000 km.

Above this display is a replica of a ship's bridge featuring the wheel, binnacle, engine-room telegraph, radar scanner, and ship's bell from the *Monowai* and an early chart of the Waitemata Harbour. On the bridge is a ventilator from an Auckland ferryboat and a depth sounder from *HMNZS Endeavour.*

Another diorama is of the Battle of the River Plate which took place in 1939 and was the first engagement in which New Zealanders served in World War II. It shows the engagement between *HMS Ajax, HMS Achilles, HMS Exeter,* and the German Pocket Battleship *Admiral Graf Spee.* This is augmented by an extensive collection of ships in bottles painstakingly made and presented by Mr L. Ricketts of Titirangi.

Maritime displays include a model of the *mv Port Auckland,* a refrigerated cargo ship of nearly 13000 tonnes gross. It is 170 m long, 21 m wide, and has a 10 m draught. Built in 1949 by Hawthorne Leslie Ltd of Tyneside, it carries refrigerated and general cargo in addition to twelve passengers, who voyage in luxurious comfort.

The centre display is a collection of vintage diving equipment with a working air pump. This type of underwater equipment is now largely superseded by wet suits and aqualungs, which allow the diver freedom from the cumbersome weight of the earlier equipment.

Motat is also in possession of a periscope from the United States Naval submarine *Cabrilla* (SS288). This vessel received six battle stars for World War II service and was credited with having sunk about 39 000 tonnes of shipping. The periscope has a range of about 16 km and is mounted so that visitors can survey the museum area through it.

One central road transport display is the McLaren M6BGT, McLaren's first attempt at designing and building a sports saloon car. Finished late in 1969, it was used on the road in and around London, England, for nine months or so until his tragic accidental death. The chassis and all running gear is standard

Opposite page:
The Bass bus, well known to young and old.

M6B Can Am, with the addition of the glass fibre saloon top made to McLaren's design by Specialised Mouldings of London. To prevent it being sold to America and to ensure its finishing up in New Zealand as Bruce would have wished, his wife Pat, Dennis Hulme, and Phil Kerr, the current manager of McLaren Racing, clubbed together to buy the car. They arranged for its display at the Motat under the direction of Bruce's father Mr Les McLaren.

A popular event with children is the bowing of the animated 'Big Bib' Michelin Tyre Man after the style of a Queen Street display of the 1930s. Lesser road transport items are an Ashford Litter, a handcart vehicle, Auckland's first ambulance and a collection of early petrol cans and boxes used before petrol pumps became numerous.

There is a display of the 1953 Coronation coach of Queen Elizabeth II, complete with attendants and outriders, together with life-size replicas of the crown jewels.

An airline theatrette is also featured. Viewers can sit in aircraft seats and watch colour presentation of the scenic attractions served by New Zealand airlines. There is also a cut-away presenation of the fuselage of an Electra 10A of 1934, the first all-metal airliner used in New Zealand. This was built by the Lockheed Corporation and used in New Zealand by Union Airways Ltd.

Other display material in the Transport Pavilion includes model railway locomotives and carriages, a railway diorama, a collection of bicycles, penny farthings, velocipeds, tandems, and 74 flags of the world's airlines which were flown at the IATA conference held in Auckland in 1973.

Motat collection

1

2

Motat collection

1. Ship's bridge in the Marine Section of the Transport Pavilion. Every lad aspires to be the skipper.

2. Section of the Transport Pavilion looking towards the mezzanine floor, Railway Section.

1. A view of the racing car section of the Transport Pavilion. In the foreground 1908 Sizaire et Naudin, in contrast with Bruce McLaren's Cooper Climax Formula I.

Motat collection

1

S. J. Woods

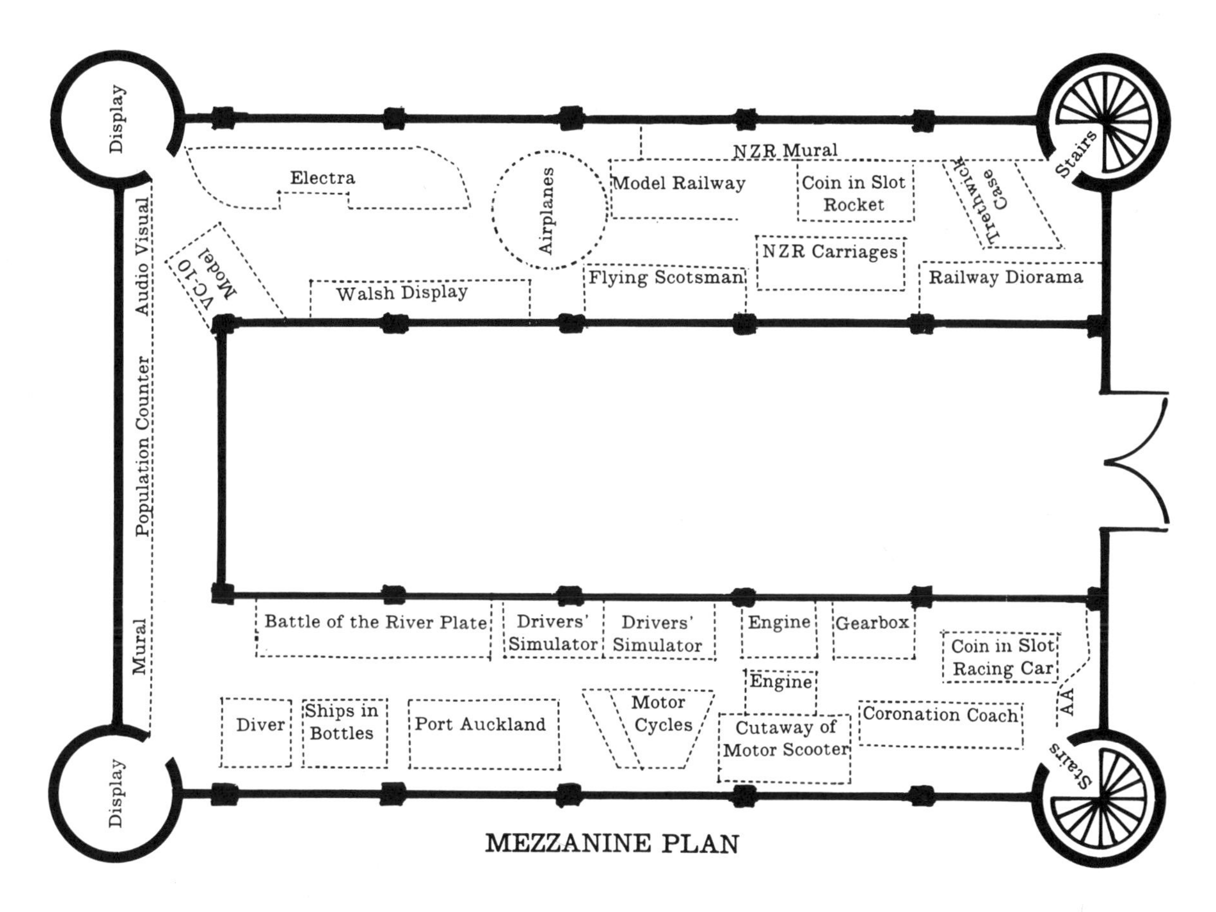

MEZZANINE PLAN

Transport Pavilion

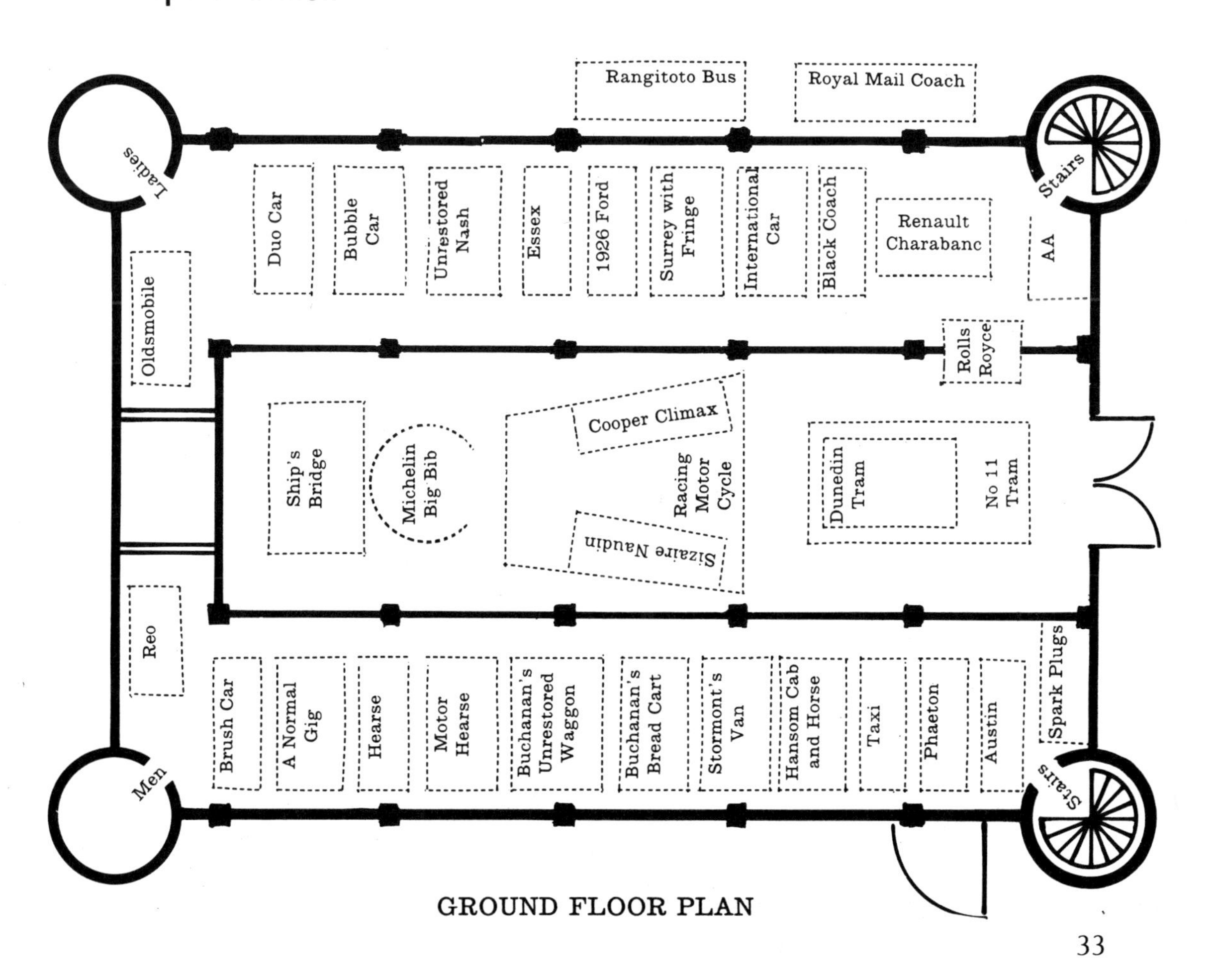

GROUND FLOOR PLAN

1

Motat collection

1. The newly arrived horse-drawn, hand-operated ladder poses for a photo in the back yard of the old Auckland Central Fire Station (now the St. John Ambulance Station). Brought to the city by the Auckland City Council in 1902 following a commission of inquiry into the fatal fire at the Grand Hotel, the ladder was a Magirus, sold under licence by Shand, Mason and Co. It was drawn for several years by two horses, then by a Kissel Kar hose tender. It was phased out of service following the introduction of a battery-driven, 26-m ladder in 1912 but was used for training purposes till well into the 1920s.

Opposite page:

The Shand Mason ladder 73 years later outside Motat's Transport Pavilion.

Fire-fighting

Fire-fighting in New Zealand has developed through several major phases. English hand-drawn and horse-drawn equipment was imported until the early part of this century. And the first motor fire appliances were mostly English, though a few American chassis were fitted with New Zealand-built bodies. Until the 1930s, heavy English pumps were the mainstay of urban fire-fighting, but from 1932, with the advent of the Ford V8 motor and chassis plus the development of New Zealand-built pumps and New World bodies, a distinctive local tradition arose.

Motat's fire-fighting collection covers the years from 1860 to 1953 in terms of major exhibits, with some later examples of small equipment.

Although Motat inherited a 1902 horse-drawn extension ladder in 1964, it was not until 1967, when a Dennis fire appliance of 1925 was acquired, that a Fire Section was formed within the museum. This was founded by three young Auckland firemen who were anxious to preserve some of the history of their profession. The subsequent years of strenous work by the section has brought recognition and acceptance by the New Zealand Fire Service.

The Fire Station at the museum was officially opened in May, 1974, by the Chairman of the New Zealand Fire Services Council, to coincide with the centenary of the Auckland Metropolitan Fire Board. It occupies a converted and renovated World War I army building, the eastern half of which has been made into a three-bay appliance room with the front of each bay fitted with a four-leaf folding door from the former Mt Eden Fire Station.

The watchroom incorporates an alarm panel from the fire station at Hobsonville RNZAF Base, and a small attic gives access to a wooden pole from another station. Also on display in the watchroom are sets of old street fire alarm codes written on blinds. These are from the Remuera and Avondale stations and are shown together with firemen's service certificates and other items.

The western half of the building is a large hall housing horse-drawn appliances, a wall display of extinguishers, display cases with uniforms, helmets, medals, breathing apparatus, and water way equipment together

F. Neill

with a collection of photographs depicting firemen and fire-fighting in New Zealand over a period of a century. One of the features of the uniform collection is the helmet, axe, and pouch worn by the first Auckland Chief Fire Officer, Asher Asher (1857-1874), and the whistle, long service badges, and decorations of the second Fire Chief, John Hughes (1874-1899).

Motat has representative exhibits from all eras of New Zealand fire-fighting: horse-drawn, motor, pre-World War I, post war, and modern.

The 1868 horse-drawn Shand, Mason & Co. manual pump is typical of the first fire engine brought to New Zealand. It has an internal vertical plunger pump activated manually by horizontal rocking arms on each side. This machine was used for an incredible 86 years, by the Reefton Volunteer Fire Brigade from 1869 to 1937, by the Auckland Metropolitan Fire Board for the Auckland City Centennial in 1940, and from then until 1955 by the Wellsford Volunteer Fire Brigade.

Another Shand, Mason exhibit is the 1902, 20-m, horse-drawn Magirus three-section extension ladder. There were only three of these machines imported. This one, supplied from England to a German design, saw operational service with the Auckland Fire Board from 1902 to 1913 before it was retired to use for drill and display purposes.

The 1907 Merryweather Chemical Hose Reel Tender at Motat is thought to be the oldest self-propelled fire appliance in the Southern Hemisphere. This engine served all its operational life in New South Wales, Australia, but is typical of those used in Wellington, Christchurch, and Dunedin.

The 1925 Dennis pump appliance is typical of the forty or so pump and hose reel tenders of this make which were brought into New Zealand in the period before 1930. It was used by the Hawera Fire Board until 1954 and then at Okaiawa until 1963. Of the same period are the 1926 and 1927 Halley Simonis pump appliance and hose reel tenders. In the late 1920s Henry Simonis & Co. supplied pumps and hose reels for use on a variety of British vehicles including the Glasgow-built Halley chassis with a six-cylinder Taylor or Dorman motor. The Taylor was built under licence to the Renault Co. of France.

The Auckland Fire Board owned six of the

seven of these appliances in the country. They served at various times at Central, Western District, Mt Albert, Remuera, and Onehunga stations. One was later sold to Henderson and another to Howick. Motat has two more or less complete chassis and motors plus parts from these vehicles, and it is hoped that at least one appliance can be restored.

The oldest modern type appliance is the 1932 Morris Commercial hose reel tender. In 1932 the Auckland Fire Board imported three 1525-kg Morris Commercial chassis for the addition of a locally built hose locker body and a 135-litre soda acid extinguisher (removed from earlier Kissel Kar or Thornycroft appliances).

2

F. Neill

Alexander Turnbull Library

1

3

F. Neill

Opposite page:

The 1868 Shand Mason manual pump with a hose cart, in the forecourt of the Pioneer Village. Originally named *Excelsior,* this manual started service in the South Island West Coast town of Reefton, where it served for an incredible 83 years before coming to Auckland in 1940 to take part in centennial celebrations. The Northland town of Wellsford purchased it and used it as the community's sole fire engine until 1955, when it failed to contain a fire which destroyed most of the business centre of the town.

1. A 1906 Merryweather motor fire engine and crew outside the Thorndon (Wellington) Fire Station before 1910. This is a similar appliance to the one held by Motat.

2. An 1886 Merryweather steam fire engine is given a trial run through the museum grounds prior to taking part in the Auckland Metropolitan Fire Board's Centennial Parade in May, 1974. Regrettably, this is the only fire engine of its type that Motat has so far been able to obtain, being loaned by the Ferrymead Trust in Christchurch.

3. 'Ah, those were the days'. A pensive Fire Section member tries to visualise what a drive through the streets of Sydney would have been like in 1907 in what was one of Australia's first motorised fire engines. This vehicle, with non-original Dennis body work, is probably one of the oldest motorised fire engines in existence and when restored completely will be one of the museum's prize exhibits.

1

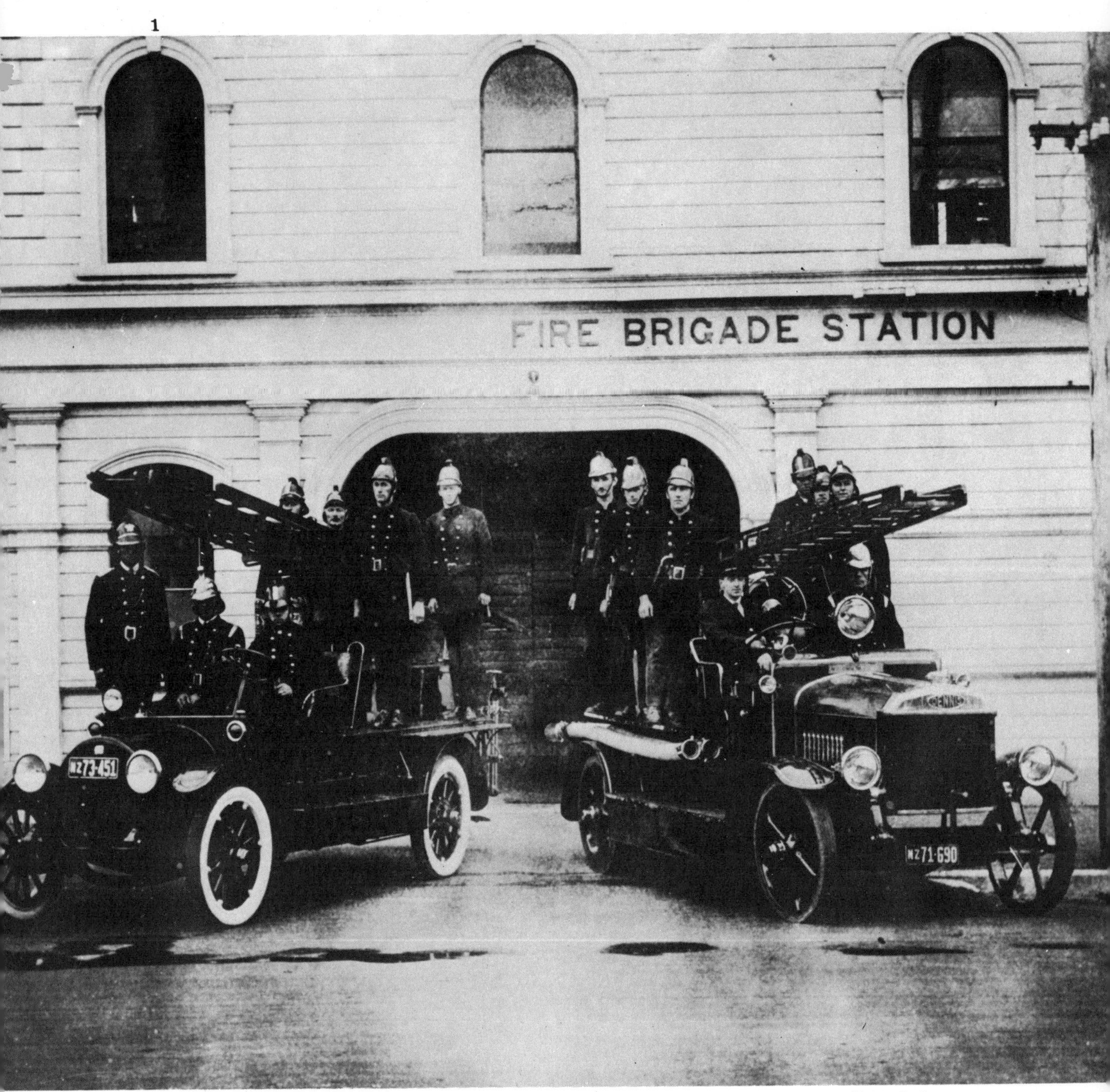

Motat collection

1. Motat's 1925 Dennis pump shown in service outside the Hawera Fire Station in 1926, along with an early model Buick hose tender.

1. New Zealand's first fully-enclosed fire appliance, a 1942 Ford V8, with Colonial Motor Co. 1800-litre-per-minute rear-mounted pump, photographed in service in December, 1972.

2. New Zealand's only 1948 Leyland appliance with a 30-m Merryweather turn-table ladder, in service at Parnell, Auckland, in March, 1966.

3. This 1955 Ford V8 F750 'Big Job', with Marmon Herrington All-Wheel Drive, was built in New Zealand as an airport crash tender. It is fitted with an 1800-litre water tank, 280-litre foam compound tank, 2250-litre-per-minute Colonial Motor Co. rear-mounted pump, and a foam monitor.

2

F. Neill

1

F. Neill

3

F. Neill

1. Fully restored at Motat since 1967, this 1925 Dennis fitted with a 100-litre-per-minute rear-mounted pump stands proudly in the museum Pioneer Village. It is typical of the British-built appliances which formed the mainstay of New Zealand's urban fire-fighting plant from 1910 to 1930.

2. The 1902 Shand, Mason & Co. Magirus 20-m extension ladder, fully restored and paraded with its horse power 72 years after its arrival and commissioning in Auckland.

F. Neill

1

2

E. Neill

F. Neill

1. *Indestructible*, a 1932 Morris Commercial Chemical Tender, leaves a private collection for its final resting place at Motat, October 1974. This fire engine served with the Auckland Fire Brigade's Parnell Fire Station before being bought by the Glen Eden Volunteer Brigade in 1948. It served at Glen Eden until 1963 when it was nearly destroyed by fire when the station was gutted in December of that year. The local kindergarten was given the old engine, and there it suffered until in 1973 it was acquired for a private collection, whence it was driven under its own power to Motat. It is one of two known survivors of the many fire engines which relied on a chemical reaction in a large tank to produce a jet of water from the appliance's hose reel.

1

In 1932 the Colonial Motor Co. Ltd and its associate, Standard Motor Bodies Ltd, both subsidiaries of the Ford Motor Co. and based in Wellington, entered the New Zealand fire appliance manufacturing field. Using British Fordson V8 and American and Canadian Ford V8 motors and chassis, the two companies dominated the New Zealand fire appliance market for 20 years. Originally American Hale pumps were utilised, but soon Colmoco, as the company became popularly known, built its own centrifugal pump for front-, mid-, and rear-mounted positions. Motat has a number of Colmoco products, some with Standard-built bodies.

Four of these are ordinary civilian appliances, the oldest dating from 1935. This model has a 4-m wheelbase and a Colmoco 1800 l.p.m. mid-mounted pump. It is one of five of this pattern used in New Zealand and saw service with the Auckland Metropolitan Fire Board for 30 years and with the North Shore Fire Board for another 10.

The second engine is also a 1935 model and has a standard 3.3 m wheelbase and a Colmoco 1600 l.p.m. front-mounted pump. It was one of five of a design built for small-town service. It was in service for 20 years in Dargaville, followed by 14 years at Silverdale and Manly and 2 years at Kumeu. The third and fourth appliances, built in 1937 and 1939, are the standard type and feature the widely used rear-mounted pump. Both served for many years in Auckland, the 1937 Ford going to Whitianga in 1955 and the 1939 model to Silverdale in 1962.

Comolco built several thousand Ford V8-powered 1800 l.p.m. trailer-mounted pumps for war-time service in New Zealand and overseas. Motat has a 1942 restored pump which has seen post-war service with the New Zealand Forest Service's Waipa Mill and with the Pureora Forest Volunteer Fire Brigade.

One of the most interesting and unusual of the Fire Section's engines is the 1942 Ford V8 four-wheel-drive with a rear-engined, military-pattern chassis. This was built as an armoured car and rebuilt in 1946, along with 30 others, to the design of the New Zealand Forest Service, as a 3500-litre tanker with a small PTO chain-drive gear pump. The Fire Section's exhibit was used at Riverhead Forest from 1946 to 1969. The museum's Aviation Section has a sister vehicle which served at Tairua Forest from 1946 to 1970.

The most modern of the museum's appliances is a 1951 Ford V8 Marmon Herrington 'all wheel drive' airport crash/fire tender with a Colmoco 2200 l.p.m. rear-mounted pump, 1800-litre water tank, and 300-litre foam tank. This was the prototype of 20 such appliances built for the Royal New Zealand Air Force and the Civil Aviation Administration, and it saw its final service at the Crash/Fire School at Christchurch International Airport.

Motat also has a modernised Fargo appliance dating from 1944. This was one of about 25 Fargo T118 chassis with 6-cylinder motors brought into New Zealand for the Government National Service Department but deployed in high fire-risk areas and manned by local-authority fire-fighters. The usual design, a combination pump/hoselayer and called a Water Unit, incorporated a Colmoco 1800 l.p.m. pump with its own Ford V8 motor, skid-mounted across the chassis behind the cab and followed up with a New World hose locker body suitable for laying a mile of 9.5-cm Landline canvas hose.

After the war, ten Fargos were rebuilt with four-door, six-man cabs, most having their pumps moved to the rear of the appliance and fitted to a power take-off. Motat's Fargo commenced service with a 3000 l.p.m. Leyland pump with a Leyland Cub motor, skid-mounted at the rear. It served with the Auckland and Henderson Brigades. The section also has a Dennis heavy trailer pump, one of twenty imported during World War II for use with Fargo pump/hoselayers. This particular pump was used by the Wanganui and Mangaweka Fire Brigades.

In addition to the horse- and motor-driven pumps, Motat has a collection of portable pumps. The oldest of these is an Eclipse manual pump, built by W. H. Price in Christchurch in 1896.

3

F. Neill

1. 2. This 1935 Ford V8 with a 1600-l.p.m., front-mounted pump started service with the Dargaville Fire Brigade and was not replaced until 1956. It was acquired for the Silverdale district by public subscription, rebuilt, and used there until 1970. It then served the Kumeu district until 1972, when it passed to the museum.

3, A good example of a New Zealand-designed, small, fast fire engine for use in rural areas. Presented to Motat by the Papatoetoe Fire Brigade, this 1955 Ford V8 is still in service, being used as a fire engine within the museum.

1 2

F. Neill

E. Neill

2

F. Neill

1. *The Rattler*. Shown here at the time of its commissioning in March, 1936, this 1935 Ford V8 was the first of its make to see service in Auckland. It was extremely fast, serving as the 'Flyer' at the central station for nearly 20 years. Nicknamed *The Rattler*, the appliance served in Auckland until 1965 when it was purchased by the North Shore Fire Board, who donated it to Motat in 1974.

2. Forestry Fire Engine
At the end of World War II the New Zealand Forest Service acquired a number of Ford V8 armoured car chassis. The Service constructed water tanks of 36 370-litre capacity on and around the chassis, fitted a 450-l.p.m. gear pump, and issued them to forests around the country. The engine shown above served at Riverhead Forest until 1970, when it was acquired by Motat.

3. The Auckland Metropolitan Fire Board purchased 15 Ford fire engines in all. All except three followed the style of the engine pictured. This particular machine entered service in 1939 and served at the St. Heliers and Otahuhu Fire Stations until transferred to Silverdale in 1962. It was donated to Motat in 1973.

F. Neill 3

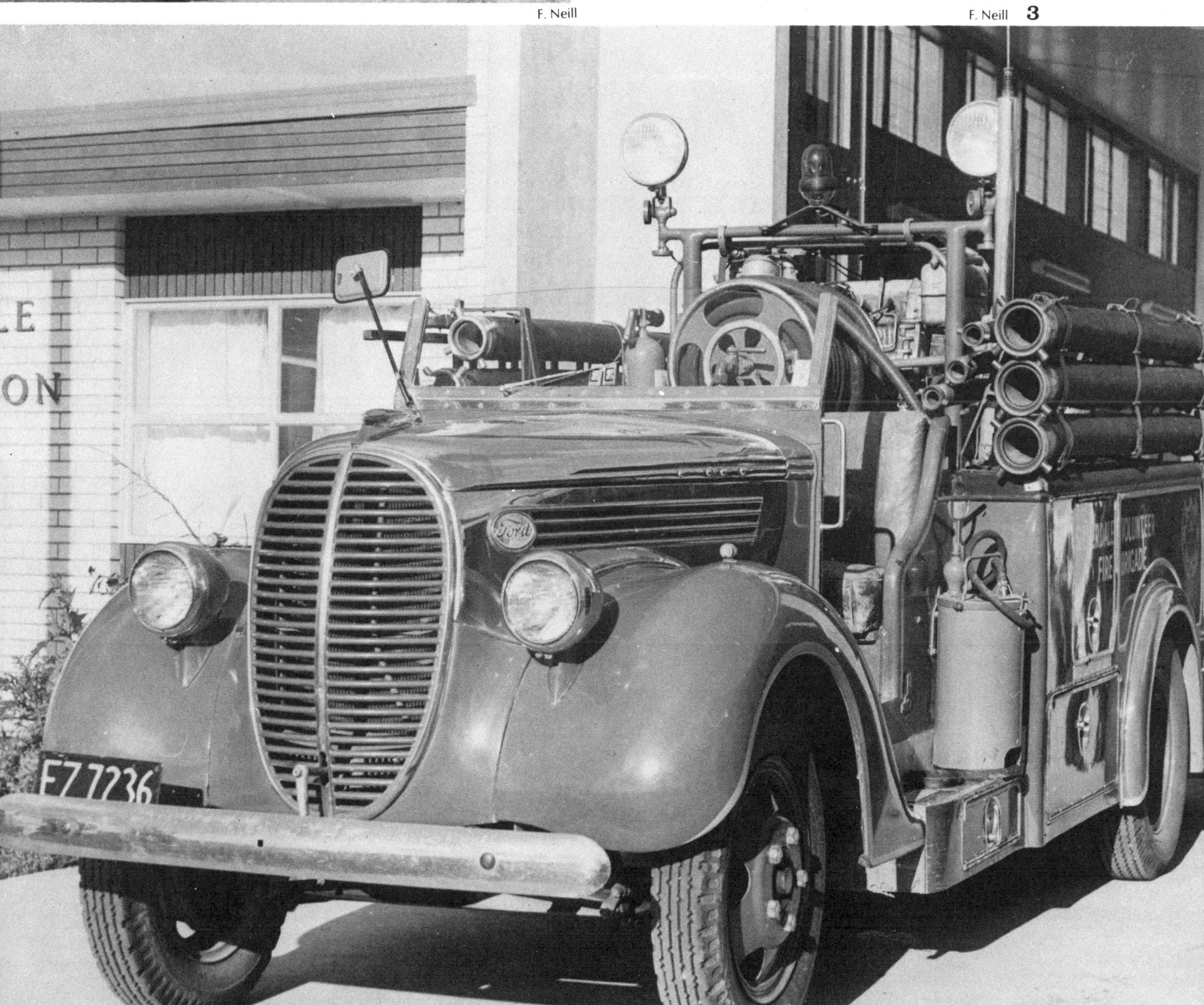

WESTERN SPRINGS
TRAMWAY
TRAMS LEAVE THIS ZONE FOR
A RETURN TRIP THROUGH
THE MUSEUM GROUNDS
MON - FRI BY ARRANGEMENT
SAT - SUN EVERY 10 15 MINS
PUBLIC HOLIDAYS AS TIMETABLE
- FARES -
ADULTS & CHILDREN 5c
CHILDREN UNDER 5 FREE
- NOTICE -
RAILWAY
SPECIAL FARE 6D
Go well-Go Shell

TRAMWAYS

New Zealand's tramcar era began in Nelson in 1862 with a passenger-carrying horse-drawn tram running from Trafalgar St to the Port. It ceased in Wellington in 1964 when the last electric tram was withdrawn from service.

At first, trams were pulled by one or more horses, depending on the hills to be encountered. In 1878 steam was introduced in Wellington as a means of propellant. Steam services were taken up in Dunedin in 1879, Christchurch in 1880, and the North Shore of Auckland in 1910. Horses were still used throughout the steam era but were gradually replaced by electric trams.

The Dunedin and Roslyn Tramway Company introduced the first electric tram service in 1900, and this was followed by the complete electrification of Auckland services in 1902. Dunedin followed suit in 1903, Wellington in 1904, and Christchurch in 1905. Other large provincial towns also adopted electric services: Wanganui in 1908, Invercargill in 1912, Napier and Gisborne in 1913, and New Plymouth in 1916.

Cable tramways also operated in a number of places at this time. The Dunedin suburb of Roslyn was the first, establishing its service as early as 1881. Wellington installed a cable tram service to Kelburn and it still operates today.

The tramway exhibits at Motat represent a hundred years of urban rail transport in New Zealand. The preservation of New Zealand trams began when the Auckland Transport Board (now taken over by the Auckland Regional Authority) presented tramcar No. 11 to the Auckland City Council in 1952. The country's oldest electric tramcar, it was the first of 43 to be assembled in 1902 and later decorated and run over the Auckland system to commemorate fifty years of electric trams in Auckland. The Sir John Logan Campbell Trust donated a shelter for the tram, and it was housed at the Auckland Zoological Park until presented to the museum.

In service for 50 years, it is a double-truck combination type tramcar with a centre saloon and a semi-enclosed compartment and driver's platform at each end. Built in England by Brush Electrical Engineering Company, it is powered by two 37 kW motors. At 12 metres long, it seats 50.

Graham C. Stewart

1

2

Graham C. Stewart

1. Wanganui Steam tram No. 100 puffing along Victoria Avenue hauling two convertible type trailers during the celebrations to farewell the Wanganui trams in 1950. No. 100 and convertible trailer No. 21 are both owned by Motat and when completely restored will be used on special occasions to carry passengers on the museum tramway.

2. A close-up view of the 1891 Baldwin ex Sydney Steam tram No. 100 as she appeared during the Wanganui tram farewell celebrations in 1950.

Opposite page:

Tram No. 253 in passenger service on the museum tramway. The concrete safety zone is a replica Auckland type.

Cobb

Tram preservation did not start in earnest until the Auckland system was giving way to trolley and diesel buses in 1956 and a tramway historian, G. C. Stewart, suggested to the Auckland Transport Board that, as tram No. 11 was preserved, it would be logical to preserve for posterity the most modern tramcar, No. 253. When nobody was willing to take care of this item, Mr Stewart formed a trust to preserve it, and this developed in 1957 into a society called the Old-Time Transport Preservation League.

The items acquired for the society were housed on the Matakohe property of the League's president, M. D. Sterling, now Keeper of Exhibits at Motat. They were later presented to Motat, to form the basis of today's Tramway and Agricultural Sections.

At first the Tramway division was part of the Railway Section, but in 1966 the trams were reorganised into a separate section which now operates all the tramway system. From the start it had been envisaged that the trams would operate on track laid within the Western Springs complex. The job of laying tracks was begun in 1964. These had to be dual gauge so that they could carry tramcars from various towns.

In spite of great difficulty with rocky outcrops and extensive manual labour, the track was completed to be extended to take trams as far as the nearby zoo. Strict authenticity has been preserved by using original poles, painted in the traditional green, to carry the overhead wires. Where possible the original items have been used, having been rescued from the demolition of the Auckland system. The present tramway receives its 550-volt direct current power from the trolley bus system which operates on Great North Road.

The first Motat trams to operate, No. 253 and No. 257, were launched on the new tracks by the then Mayor of Auckland, Dr R. G. McElroy, and Minister of Transport, the Hon. P. Gordon, in December, 1967, and today there is a daily service complete with motorman and conductor or conductresses, all as it used to be and complying with Government regulations.

The Motat exhibits include a great variety of tramcars and trailers, including a horse tram, No. 4, of 1883. This quaint little vehicle was built by the Jones Car Company of Schenectady, New York. For over 70 years it served as a trailer on the Mornington cable car line in Dunedin, the only vehicle to survive a depot fire in 1902. It finished service as trailer car No. 107. After the Mornington cable car service was abolished in 1957, it was sold privately for a garden shed but was rescued and donated to the Old-Time Transport Preservation League. Although used as a trailer all its life it was actually a horse tram of standard type and has been restored in this form.

The steam tram No. 100 was originally built for the tramways in Sydney but was brought to New Zealand in 1911 to be used for construction work by the contractors building the Gonville and Castlecliffe electric tramway at Wanganui. After the completion of this job it remained unused until 1920, when it formed part of a skeleton service pulling trailers during a power crisis. Its last appearance was in 1950 when it was used to pull trailers during the celebrations to mark the end of the Wanganui electric tram service.

A few years later it was saved from the scrap merchant by G. C. Stewart and P. J. Mellor, who donated it to the Old-Time Transport Preservation League. Like all steam trams of the period, it has a cover in the form of a carriage to disguise the boiler and funnel and to prevent horse traffic from being frightened. It was built by the Baldwin Locomotive Works in Philadelphia in 1891 and develops 11 kW. It has a 0-4-0 engine with a 143.5-cm gauge and a laden weight of 14 tonnes. This tram is very similar to the locomotives used on the North Shore of Auckland from 1910 to 1927.

Motat collection

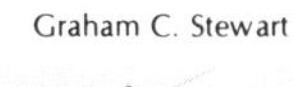
Graham C. Stewart

1

Graham C. Stewart

2

Graham C. Stewart

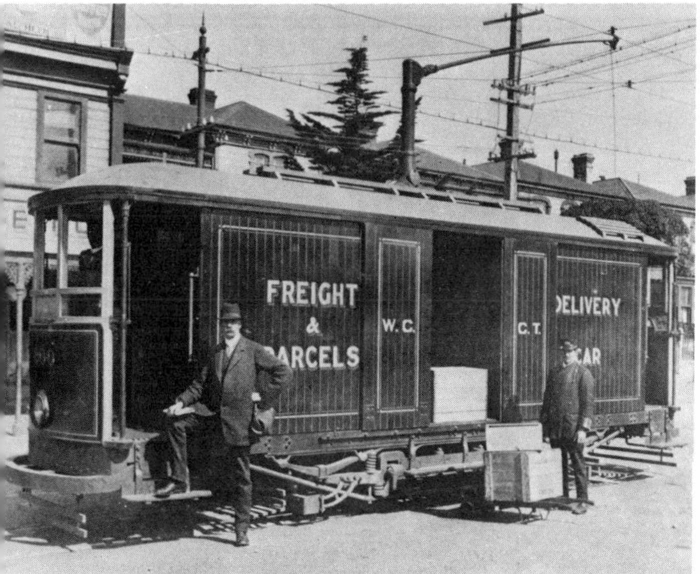

3

Opposite page:

Wanganui Electric Car No. 10, a Californian combination type built in 1912, was obtained because it is typical of trams on the smaller provincial tramway systems. It is an interesting type with its clerestory roof windows, centre saloon, and open seating accommodation at each end.

1. Another form of urban electric transport, the trolley bus is also represented at Motat. Two of Auckland's first four trolley buses, Nos. 1 and 3, are preserved at Motat. Built in 1938 on Leyland chassis with Metropolitan Vickers electrical equipment and locally built bodies, these buses were used solely on the Inner city loop that provides a free passenger service to a well known Auckland Departmental Store, The Farmers Trading Coy.

2. Motat's oldest urban transport exhibit, the 1883 tram No. 4. This vehicle, which was a standard type horse tram of the era, spent all its public life as a cable car trailer shown on the Mornington line in Dunedin.

3. In the days before the prominence of the motor lorry, many tramways used their systems for carrying freight as well as passengers. Motat is fortunate in having Wellington Freight Car No. 301 which in later years, when the freight-carrying business died, was converted to a rail grinder. It is therefore useful as a works car as well as an exhibit. The illustration shows its sister car No. 300 in service in Wellington.

4. An early photo taken about 1905 shows Motat's oldest electric tram, No. 11, travelling up Queen Street, Auckland. It is interesting to note the motorman's platform, which is completely open except for the front dash panel, and also the row of centre poles complete with side-arm brackets, and the ornamental wrought ironwork for suspending the trolley wire.

4

Graham C. Stewart

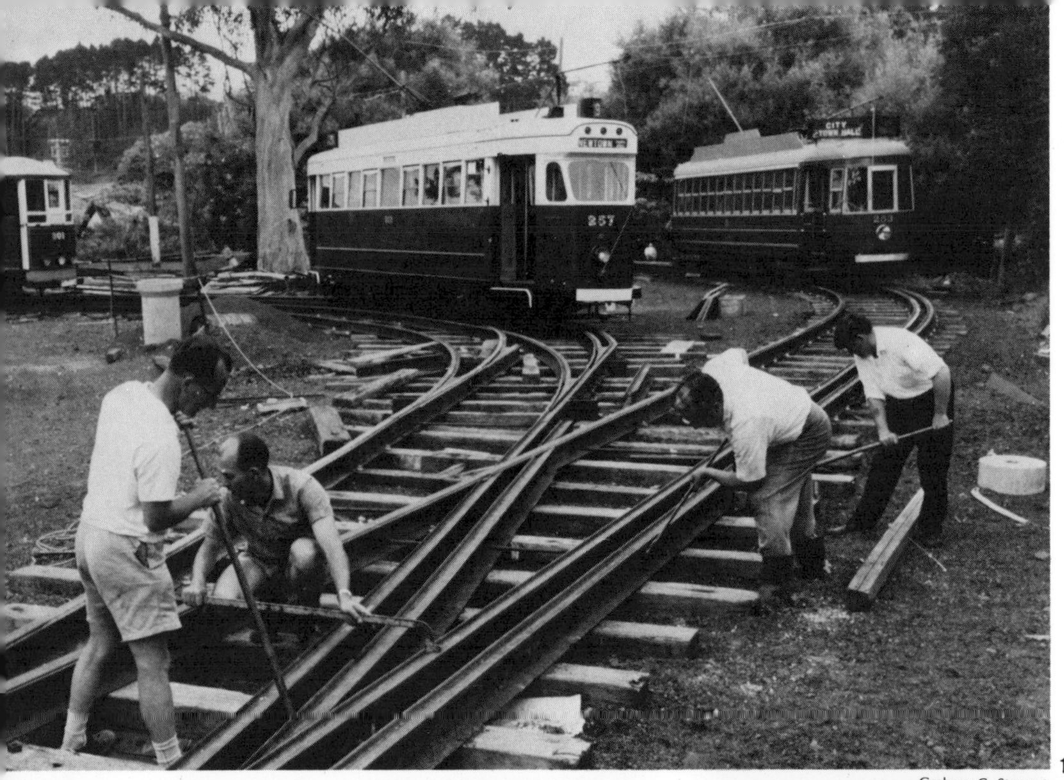

1

Graham C. Stewart

2

Motat collection

3

Motat collection

1. This picture shows early track construction being carried out by Tramway Division members. Having to lay dual-gauge track made the job more complex, especially at points and crossings, but this was necessary because the Wellington trams are gauge 121.9cm and the Auckland and Wanganui trams standard gauge 143.5cm. In the background are Wellington Freight car No. 301, Wellington Fiducia car No. 257, and Auckland Streamliner Car No. 253.

2. No. 253 and No. 11 — Auckland's last and first type cars respectively, moving away from 'Sterling' terminus. In the foreground is a side-arm bracket trolley-wire support made of totara and dating from 1902.

3. 1902 Auckland tram No. 11 leaving the lower terminus in 1952 Golden Jubilee livery. It was used to celebrate 50 years of electric tramway operation in Auckland.

Graham C. Stewart

Motat collection

Of the three types of tramcar which began the central Auckland Service in 1902, Motat has No. 44. Actually built in 1906, this is a copy of the type A four-wheel car originally built by Brush Electrical Engineering Company. From 1908 until 1922, 99 type M cars were built in Auckland, and Motat's No. 89 is one of these. No. 203, type N, was built in 1926 and is representative of the trams which serviced Auckland between 1925 and 1956.

Tram No. 253 of the Auckland Transport Board is one of six cars of the 1937 streamline class and was the fastest and most modern tram in Auckland — thus to become known as *Queen Mary*. It was built by the Board to the standard pattern of driver's cabs and large single saloon seating 52. Nearly 14 m long, it is powered by four 26-kW motors. After serving from 1940 to 1956, it was presented by the Board to the Old-Time Preservation League in 1957 and by them to Motat in 1960.

Tram No. 248, which saw service from 1938 to 1956, is similar to No. 253. In 1956 it took part in the closing ceremony for the Auckland tramway and later was the last tram to travel through Auckland streets. It was bought to be used as a glasshouse and was later presented to Motat.

No. 244 of the Wellington City Corporation Tramways Department was one of the capital's most modern trams. It was presented to Motat by Shell Oil New Zealand Ltd in 1964. With a semi-streamlined body, it is one of a series of 24 of Fiducia type made by the Corporation. It is 12 m long with 31 upholstered seats and is powered by two BTH type 502, 39-kW, 500-volt motors. It commenced service in 1939 and was withdrawn in 1964.

A similar tramcar, No. 252, started work in 1940. It was purchased by Motat at the end of its last year, 1964, the last public service tram to operate in the country. The same year, Shell Oil presented Motat with No. 257, another Wellington Fiducia type tram. This one had not been completed until 1950 because of war-time shortages of parts.

1. View of the Queen Street-Victoria Street intersection in the early 1930s, when Auckland's main form of public transport was the tramcar. Note the people waiting for a tram on the far track safety zone.

2. Tram rides are a popular feature of Motat's activities. They are a new experience for the young and bring back nostalgic memories for the old. This view is typical of the number of people taking rides during a Live Weekend.

No. 135 of the Wellington City Corporation was one of 79 introduced after 1914 and the most common type on Wellington streets. It originally had an open central compartment between two saloons, but this was partially closed in during the 1930s. The car is nearly 12 m long and seats 35. It was in service from 1921 to 1957.

Wellington tram No. 301 was one of two made by the City Tramways and introduced in 1911. It is a small four-wheel truck freight car nearly 8.5 m long, used at first for goods delivery and later as a track-grinding vehicle.

No. 47 is the oldest Wellington tram in the museum, first operating in 1906. It is an open-top double-decker with two lower saloons, seating a total of 74. Built by Rouse and Black of Wellington it is powered by two 30 kW motors and is 11 m long and 4.3 m high.

The Auckland tram No. 17, type C, was one of six double-deckers made by Brush Electrical Engineering Company and cut down to a single-deck vehicle in 1923. It is to be restored to its original form.

Typical of the tramcars used in small towns is No. 10 from Wanganui. It was built by Boon Bros. of Christchurch in 1912 to a Californian combination design, having a centre saloon and open seating at each end.

Trailer No. 21 of the Wanganui Tramways was built to be towed behind an electric car. It is convertable type that can be fully enclosed for rainy weather otherwise run with open sides. It was obtained by Motat to be towed by steam tram No. 100.

Graham C. Stewart

1

2

Graham C. Stewart

3

Motat collection

1. Wellington Combination Car No. 135 built in 1921 was typical of that fleet from the 1920s until the streamline Fiducia class made its debut.

2. Auckland car No. 203, built in 1926, is shown in service in the early 1950s. The body of this car was obtained for future restoration to show the type of tram that dominated the Auckland system from the mid 1920s to the close of the system in 1956.

3. Members of Motat's voluntary Tramway Division apply a coat of paint to the body of car No. 203 to prevent further deterioration.

Graham C. Stewart

1

2

Motat collection

3

Graham C. Stewart

4

Graham C. Stewart

1. No. 47, the 1906 open-top double-decker from Wellington, is one of the most prized in Motat's collection.

2. Motat's trams are also used for promotional purposes, and the photo shows No. 253 carrying guests within the museum grounds to celebrate the publication of the book *The End of the Penny Section,* an historic account of early New Zealand urban transport. On this occasion Lord Montagu of Beaulieu was Guest of Honour, and he is shown here receiving an enlarged replica ticket of the bygone penny section days from Conductor Mr Barry Phillips. On the left of Lord Montagu is Mr Graham Stewart, the author of the book.

3. One of the Wellington streamline Fiducia Class trams No. 257, in passenger service, passes the Pioneer Village. Motat also has Nos. 244 and 252 of the same class.

4. No. 252 is an historic Wellington tram. Pictured here is the Wellington last-tram ceremonial run of May 2, 1966, with No. 252 suitably draped with flags and bunting. As Wellington was the last city in New Zealand to withdraw trams, this car was the last in the country to run in public passenger service.

W. W. Stewart

Clearing the lower area.

RAILWAYS

Building a 107-cm-gauge railway is a considerable undertaking even when, as with Motat, only a few hundred yards of track are involved. The project was begun soon after the establishment of the museum by the Tramway Section, the Auckland Metropolitan Model Railway Club, the Railway Enthusiasts' Society, and the Bush Tramway Club.

A few feet of track, a signal box, an unserviceable locomotive *Bertha*, an equally unserviceable ex-New Zealand Railways D class locomotive, and a Fell brake van from the world-famous Rimutaka Incline comprised the lowly beginnings of one of the largest and most heavily equipped sections of the museum.

The Tramway Section soon became autonomous and developed independently, as did the Model Railway Club. The Railway Enthusiasts' Society shifted their headquarters to establish their own Glenbrook Vintage Railway, and the Bush Tramway Club was left with the task of establishing the Railway Section, equipped only with two locomotives that would not work and no track to run them on anyway.

However, by 1968 the swamp behind the museum buildings was filled, a plan for development was drawn up, a rock outcrop was removed, and two parallel tracks over 90 m long were laid and ballasted. The first rolling stock was placed on the new railway in June, 1969, and shortly afterwards the 0-4-0 Barclay steam locomotive and Hudswell Clarke diesel arrived. The newly constructed platform accommodated hundreds of visitors for the initial steaming of the Barclay in 1970.

These developments were soon followed by the construction of a cross-over between the two tracks to allow the transfer of vehicles from one track to the other. A 1910, 9000-litre, wooden water vat was also re-erected. The first stage of the locomotive overhaul pit was completed in 1972.

Workers dismantled the Waitakere station building which had been purchased from NZR for $10. The cost of transporting it by truck to Motat and re-erecting it were met by the Avondale Lions Club, who organised a golf tournament which raised some $600.

1

W. W. Stewart

W. W. Stewart

2

1. The first public steaming of the Barclay and opening of the lower area railway.

2. Waitakere Station in its heyday.

3. Motat's station, dismantled and ready to move.

3

Motat collection

1

The lean-to portion of the station had been built in 1879 as a general waiting room. It was also used as a goods shed, ticket lobby, produce stall, and bicycle shed. In 1880, a year before passenger trains began to run to Waitakere , as it was then called, the larger portion of the building was constructed at a cost of £60 for use as a Post Office.

This arrangement continued until 1911-12, when the Post Office was shifted into the general store a few yards away. The vacated building then became the ladies' waiting room. During the year 1912, installation of Tylers tablet machines for signalling resulted in a third of the ladies' waiting room being taken for this purpose. Sundry doors were added, a ticket window, partitioning in the end of the shelter, and fire buckets. The installation of electricity in 1928 was the final major event in the life of Waitakere Station.

The shifting of the building presented many problems, not the least of which was the removal of hand-wrought nails from the 93-year-old solid heart timber.

Now at Motat, the former ladies' waiting room exhibits a comprehensive display of NZR signalling equipment and a replica of a 1920 Station Master's Office. The tablet machine is now in the Railway Section Office. The lean-to shelter shed has been restored as an old-time waiting room.

The station is the nucleus of a complex to include an operating signal box, goods shed, stock yard, locomotive shed, semaphore signals, footbridge, and a railway hotel. The station was officially opened in 1973 by the then Minister of Railways, the Hon. T. M. McGuigan, who presented the Railways Section with an 1878 locomotive bell.

Opposite page:

1. Hon. T. M. MGuigan (former Minister of Railways) is assisted by Mr W. W. Stewart (noted railway artist, photographer, and historian) in cutting the ribbon to open the restored Waitakere Station, December 15, 1973; Mr E. P. Salmon, Chairman of Motat's Trustees, looks on.

2. An artist's impression of what is rapidly becoming a reality — the final development of the operating railway.

Motat collection

1

Motat collection

4

2

Motat collection

Motat collection

5

3

Motat collection

3. Part of NZR signalling display in the restored Waitakere Station, showing Tylers train tablet instruments.

4. F185 locomotive being filled from the 1910 wooden water vat.

5. The Barclay locomotive arrives at Motat. All railway equipment must be transported to the museum by road as there is no rail access.

The main locomotive exhibits at the museum are:

K900, a 4-8-4 steam locomotive (1932).

Ab832, a 4-6-2 steam locomotive (1925).

F180, an 0-6-0 saddle-tank steam locomotive. (1874).

L207, a 2-4-0 side-tank steam locomotive (1877).

A Haig class 0-6-0 side-tank steam locomotive (1918).

A Barclay 0-4-0 saddle-tank steam locomotive (1912).

A diesel 0-4-0 locomotive (1936).

A Duetz/Lenz petrol locomotive.

An Orenstein & Koppel 0-4-0 well-tank locomotive (1904).

Ww491, a 4-6-4 side-tank locomotive (1912).

1

1. A reality — an operating railway, 1975.

2. K900 on show at Pacific Steel Ltd premises at Otahuhu before being moved to Motat.

2

Motat collection

The largest and most impressive of the section's engines is K900, built at the NZR Hutt Workshops in 1932. Thirty K-class engines were built, followed by 35 Ka and 6 Kb, which were modified versions of the original type. K900 was the class leader. The K-class boiler pressure was 35.7 kg per sq.cm and the tractive effort about 14000 kg. With a total weight of 137 tonnes (engine and tender), they were the largest non-articulated locomotives in NZR service. The only larger ones were three short-lived and unsuccessful Garrett engines.

Probably the most travelled locomotive at the museum is the 4-6-2 steam locomotive Ab832, which during its lifetime steamed in excess of a million kilometres. It was built by the North British Locomotive Co., Scotland, in 1925 and was one of 141 constructed overseas and in New Zealand. Motat's was the last steam locomotive to operate in regular service in the North Island. It has a total weight of 85 tonnes, a boiler pressure of just over 32 kg per sq. cm, and a tractive effort of 9000 kg. With its light axle loading this was a versatile engine.

F180 was built by the Yorkshire Engine Company in England in 1874. Eighty-eight of these engines were built by various manufacturers, and a number were named after characters from Sir Walter Scott's novels. F180 bears the name *Meg Merrilies.* It is being maintained in good condition as a static exhibit pending the day it may be returned to steam. After retirement from the NZR, many such locomotives worked solidly for timber mills, freezing works, and so on, until forced into a second retirement by the diesel.

W. W. Stewart

1

1. F180 *Meg Merrilies* restored to her former glory.

2. Maid of all work, the commonest class of steam locomotive used on New Zealand Railways, the Ab.

2

W. W. Stewart

The side-tank steam locomotive, L207, was built by the Avonside Engine Company in 1877 and was used on construction trains on the North Island Main Trunk Railway. It hauled the first through-train over a portion of the new track in spite of a total weight of only 19 tonnes, a tractive effort of only 2 300 kg, and a boiler pressure of just over 23 kg per sq. cm. This engine saw service on the industrial railway of Wilsons Portland Cement Company at Whangarei and was finally donated to Motat by the company in 1971.

The Haig class 0-6-0 side tank steamer was built by Kerr Stuart & Company in 1926 and is of a design supplied to the Canadian Forestry Service in 1918. It is fitted with outside Walschaert's valve gear, the only locomotive with this equipment to have actually operated at Motat. This and a sister engine were purchased new and operated by Kempthorne Prosser & Co. Limited at their Otahuhu and Dunedin works. On their retirement the engines were donated respectively to Motat and to the Ocean Beach Railway at Dunedin.

The 0-4-0 Barclay saddle-tank steam locomotive was built by Andrew Barclay & Sons, Scotland, in 1912 and was the first locomotive to steam on the extended museum track. One of its operators was the Auckland Gas Company and the engine was a familiar sight hauling coal from the wharves to their works. The next owner was Pukemiro Collieries who, at the time of closing their mines, donated the engine and a carriage to Motat.

The diesel 0-4-0 was built by Hudswell Clarke of Leeds, England, in 1936. It is powered by a four-cylinder Paxman Ricardo diesel engine with a chain drive to the axles. This is started with the aid of a Briggs and Stratton auxiliary motor turning the flywheel of the diesel engine. This locomotive is ideal for yard duties and shunting, particularly on short headshunts. It has the advantage of an overall length of only 3.5 m.

Very little is known of the Deutz or Lenz single-cylinder petrol-engined locomotive. A World War I photograph shows it operating with the New Zealand Railway Group in Western Samoa. It was, at that time, built to a gauge smaller than 107 cm. It was donated to Motat by Mr W. S. Miller, who had used it at various business locations including the mercury mine at Puhi Puhi in Northland whence Motat received it. Fully restored and now affectionately named *Lizzie*, it can frequently be seen in operation, large flywheels spinning, jetting along at a top speed of 8 km/h.

Bertha, as the name would suggest, was built in Germany in 1904. It is an 0-4-0 well-tank locomotive and was built to a gauge of 60 cm. A 1918 conversion brought the gauge up to 107 cm. This 5-tonne engine was donated by Wilsons Portland Cement Company to the Old-Time Transport Preservation League of Matakohe which in turn gave it to Motat.

The side-tank locomotive Ww491, a 4-6-4 built in 1912 at the New Zealand Railway Workshops, Hillside, had an in-service weight of nearly 52 tonnes. This engine was at one time used for staff training at the New Zealand Government Railway Workshops at Otahuhu. The boiler, cylinders, steam dome, and other parts have been sectionalised to allow a view of the inside workings and make an ideal exhibit for the museum.

In additon to the railway engines, the museum has three carriages of varying types and ages, the major part of an 1879 four-wheel carriage, a specially equipped brake van from the now closed 1:15 Rimutaka Incline, and a rather obscure carriage body thought to be from the very early Riverhead-Kumeu railway.

Other rolling stock includes sheep, coal, implement, and ventilated and ballast hopper wagons of classes J, L, M, Xa, and Yb respectively. There are also track gang jiggers, trolleys, semaphore signals, station gates, a locomotive water vat, platform trolleys, and many other items of station equipment.

Motat collection

W. W. Stewart

1. 0-4-0 Barclay steaming with car A851 at Pukemiro before they were donated to Motat.

2. Orenstein & Koppell locomotive operating on a 60-cm-gauge track. (From an old print.)

Model Railway Section

The association between Motat and Auckland Metropolitan Model Railway Club has flourished since the early 1960s. The club was formed in 1961 and some two years later shifted its headquarters to Motat.

A Keith Hay Community Hall building was erected in 1968 and club members completed the installation of plumbing, electricity, lining, and general finishing. Next came an extensive scenic layout for public display. A large number of skills have been utilised in the construction of this layout, and a number of new methods of modelling scenery have been incorporated together with advanced transistorised control units. The final layout will be a working model of a New Zealand railway, complete to the finest detail.

Motat collection

3. Rolling stock in action over part of the completed track.

M. Tolich

Opposite page:

An early line-up — Hudswell Clark, Kerr-Stewart, Barclay, and carriages.

Following page:

F185 steaming during a 1975 Live Weekend.

STEAM ENGINES

If it were not for the steam section, Motat would not be where it is today. It was to preserve the great beam pumping engine, a monster so large that its upper part is lost in the high ceiling and its base disappears into the depths beneath the floor, that the decision was made to establish the museum on its present Western Springs site.

The unitiated, standing in the middle of the old Engine House, may be pardoned for asking where the engine is. True, the upper part of a huge flywheel is in clear view — 6 m in diameter and weighing over 16 tonnes — but only by looking upward and tracing out the linkage of the long connecting rods to the far end of two massive, oscillating beams and by following these beyond the tops of four heavy, cast-iron columns to more rods connecting large inverted steam cylinders does the onlooker appreciate that this is indeed a mighty engine.

The beam engine was constructed by John Key & Sons of Kirkcaldy, Scotland, and is one of the last of the type originally developed to drain the Cornish tin mines. It was erected in 1877 to pump drinking water from nearby Western Springs for the needs of a rapidly growing Auckland. Working at a steady 14.5 revolutions per minute, with each of its two pumps raising nearly 320 litres of water at every turn of the cranks, the engine was capable of providing about 545 million litres per hour.

Built at a time when gas engines were few, oil and petrol engines in their infancy, and diesels unheard of, this engine, in common with most of those of the period, used steam as its working medium. This was supplied by four two-furnace boilers installed in the adjacent room, and steam at a pressure of 9.3 kg per sq.cm was fed into a 25-cm pipe.

Passing a governor-controlled speed regularity valve, the steam entered the high-pressure valve chests and thence at the appropriate times was directed to one or other end of the two smaller cylinders. Having transmitted part of its energy to the high-pressure pistons, it was led to the low-pressure valve chests, and the performance was repeated in the two larger cylinders. After this it passed at a pressure considerably below that of the atmosphere to a condenser beneath the floor. By condensing the steam

S. J. Woods

Previous page:
The great beam pumping engine gets all the attention. C. Nissen

rather than exhausting it to atmosphere, considerably more power was extracted from the steam, and hot water of pure quality was available for pumping back into the boilers.

The actual pumps, which delivered water to a reservoir in Ponsonby, were far below the floor of the Engine House and were operated by a further pair of long rods which can be seen hanging from the beams between the cranks and the columns. Only one of these pumps still remains. The other was melted down for scrap during World War II. The pumping engine was stopped finally in 1936 when more economical sources of water became available. The boilers were removed for use in other locations, two at Auckland Hospital.

In the brick boiler house the museum's engineers have erected an impressive ship's steam reciprocating engine, a small one of its type. It dates from the early 1900s and belonged to the Sydney passenger ferry *Greycliffe*. This vessel sank with a large loss of life in 1927 when it was struck by the New Zealand vessel *Tahiti* in Sydney Harbour. The engine was salvaged and brought to New Zealand and used for over 30 years at the Tirau Dairy Factory. With the electrification of the factory, the engine was presented to Motat. It has been overhauled under the direction of Mr S. Lambert and will one day be a working exhibit.

Where the expansive property of steam is used in most engines, the earliest had only a single cylinder and is described as a simple steam engine. The *Greycliffe* engine carries the story of steam a step farther than the beam engine, for it has three cylinders, all of different sizes.

Samuel Hornblower made the first compound engine in 1782 by providing a small-diameter first cylinder and a larger-diameter second cylinder, thus providing two expansions. The third cylinder of the *Greycliffe* engine makes it a triple-expansion engine. The main advantages of the triple expansion engine are a lower range of temperature in each cylinder, a more even turning motion due to the additional crank, and more efficient use of steam. Many engines with four different sizes of cylinders were built, mostly for ships, but today few vessels driven by the reciprocating type of steam engine are still in commission.

The *Greycliffe* engine is not completely original. A ship's engine is not fitted with a large flywheel — a large grooved wheel was added for use at Tirau where the dairy machinery was belt-driven. The ball-type governor fitted at the steam inlet was also added for factory use.

A ship's engine of this type must be reversible, and it can be seen that two eccentrics are provided for each cylinder, one for forward motion and one for reverse. The transmission of the movement to the steam distribution valves is by means of the famous Stephenson Link Motion.

As with all maritime steam engines, a condenser is provided for re-use of water. An air pump, lever-driven from the high-pressure unit, is fitted to maintain a vacuum in the condenser, so extracting more work from the steam. Two small ram pumps driven by the same levers were originally used for forcing make-up water into the boilers.

Next to the *Greycliffe* engine in the museum is a steam boiler made by Daniel Adamson in 1956. This has been installed for use in making a number of the exhibits come to life again. It is a single-furnace, dry-back, multitubular type and is heated with an oil-burning unit using diesel fuel. Near at hand, standing against the wall, is the connected feed water pump made by G. & J. Weir of Cathcart, Glasgow. This pump supplies water to the boiler to make up for the steam used. A similar but much larger pump has an added refinement — provision for varying the degree of expansion which the steam undergoes in the cylinder.

Motat collection

1

Motat collection

2

Motat collection

3

1. The controls of the Wallis & Steevens 10-tonne steam roller are explained to Bishop Gowing by Mr Jim Burnham.
2. Affectionately known as Rock 'n Roll, this Aveling & Porter road roller is driven by a single-cylinder oil engine.
3. The McLaren Traction Engine after conversion from locomotive type boiler to forced-circulation steam generator.

Also in the Engine House is a steam-driven donkey feed pump dating from the late nineteenth century. Instead of having the steam cylinder in line with the pump as has the Weir pump, this exhibit has a crank and a flywheel. The Weir pump is able to work without these because the valve controlling distribution of the steam to the top or bottom of the cylinder is itself steam-operated by means of an auxiliary valve. This valve is controlled by the movement of the pump-rod.

A Weir turbo-feed pump provides a contrasting means of doing the same job. In this case the driving unit is a small steam turbine which provides rotary motion for a centrifugal pump. The actuating force is obtained by making a high velocity steam jet impinge on a series of radial blades projecting from a revolving disc. The pumping action results from water being expelled by centrifugal force from radiating passages in a revolving wheel. The turbo-feed pump is capable of pumping more water at a higher pressure than the piston type of pump. This pump was presented by the University of Auckland.

Another major item on display is a Belliss & Morcom steam engine coupled to a dynamo for generating direct-current electricity. This machine dates from about 1920 and is typical of medium-speed steam engines produced in the period 1900-1940. For the technically-minded, an ingenious feature of the engine is the use of only one eccentric to serve the two cylinders. This is rendered possible by supplying steam between the rings of the high-pressure distribution valve ('inside steam') and exhausting from between the rings of the low-pressure valve ('outside steam'). Both valves are of the piston type.

A Tangye horizontal compound steam engine has been installed near the front of the display and has been direct-coupled to a triple-ram pump. This pumping engine is the North Shore brother of the big beam engine, in that it pumped water for many years from Lake Pupuke to supply residents of Milford and Takapuna. In its original setting, a second and similar pump was coupled to the other side of the engine.

The museum exhibit was moved and set up in its present position in 1968 by a team of volunteers directed by Mr B. Robert. A notable feature of this engine is the second eccentric on the high-pressure side which drives an adjustable expansion valve which regulates the point of cut-off of steam to either end of the high-pressure cylinder.

1

Motat collection

2

Motat collection

1. This view of the front of the Daniel Adamson Boiler, taken during installation, gives an idea of the size of the furnace tube and shows the uptake ends of the smoke tubes.

2. Preparing the museum grounds for tar-sealing, using Motat's Aveling & Porter (Rock 'n Roll) and the contractor's Marshall road roller.

At the front of the Steam House there is a 31-kW type H horizontal single-cylinder diesel engine, made by Ruston & Hornsby in 1932. It was presented by Mr H. P. Minchen of Rawene where it worked for many years driving a sawmill. The engine is in working order and is used to belt-drive a dynamo of early design. Alongside is an early railcar gasoline engine given by NZR.

The vertical, three-cylinder diesel engine, one of a pair from the Dunedin Sewage Pumping Station, was one of the earliest built by the Swiss firm of Sulyer Bros. It dates from 1904. (Rudolf Diesel, a German engineer, built the first successful diesel engine in 1897). This engine is a four-stroke constructed to develop 90 kW at a speed of 200 revolutions per minute. Unfortunately it lay in unpreserved storage for many years after dismantling, and it is not anticipated that it will be a working exhibit.

Although the Steam House is given over mostly to large steam engines, it is also a convenient parking-place for a very early petrol-engined tractor made by Saunderson & Gifkin of Bedford in 1902. This remarkable machine is in working order and is driven by a two-cylinder engine which gives it a speed of nearly 10 km/h.

The display in the Pump and Steam House buildings is completed by several examples of small steam engines. Of particular interest is a single-cylinder reversable engine made by A. & G. Price at Thames. Single-cylinder and twin-cylinder non-reversable engines are also exhibited.

There are also a number of models. One of these is of the first passenger engine, which ran on the Stockton & Darlington Railway in 1825. There is also a fine model of a Babcock & Wilcox water-tube boiler, fitted with economiser and superheater and fired by a travelling grate stoker. This is the type of boiler installed in many electricity-generating stations early this century.

For all the size and power of the beam engine and other exhibits around it, the two outdoor steam engines command the most attention from visitors. The first of these is the steam road roller of a fairly late design, built by Wallis & Stevens of Basingstoke, England. This machine was built about 1935 and weighs nearly 10 tonnes. It is driven by a double-cylinder steam engine with a boiler of orthodox furnace-plus-smoke-tubes design. It has a working pressure of 25 kg per sq.cm. Steam is usually raised on Live Weekends when the somewhat cumbersome vehicle may be seen trundling around the museum grounds.

The most recently acquired of the steam exhibits is another mobile monster — a steam traction engine made in 1916 by J. & H. McLaren of Leeds. It is usually parked in its own covered shed near the Transport Pavilion. This engine worked out its useful life in the Christchurch district and was presented to Motat in 1971 by Haywrights Ltd. It is a type which was used in considerable numbers by the farming community during the early part of this century, driving a threshing machine or (in partnership with a similar machine) pulling a multiple plough from one side of a paddock to the other. It was a common sight to see these monsters on country roads, drawing a train that might consist of threshing machine, caravan, and water cart, en route from one farm to another.

The engine has high- and low-pressure cylinders and a ball type governor for maintaining a constant speed while ploughing or threshing. The transmission consists of a three-speed gear arrangement followed by a differential. Reversal, as with most steam-driven vehicles, is by changing the direction of the engine's rotation.

The boiler was originally of the conventional smoke-tube type. Unfortunately, it was condemned after its first annual survey revealed cracks in the water-jacket of the firebox. The cost involved and the risks of further cracks developing made repair inadvisable, and the huge engine looked to be useful only as a static exhibit. However, the museum's Steam Engineer, Mr J. Burnham, suggested installing a boiler on the lines of those used in the White Steam Car and the Museum Director, Mr Richardson, advised the use of a Kerrick steam-cleaner unit.

A second-hand Kerrick steam generator coil was located and found to fit snugly into the original boiler shell. Instead of solid fuel a diesel fuel pump and injector were installed. A blower was needed to provide air for combustion, and a forced-circulation pump was required to pump water through the coil in which it was to be raised to a much higher temperature than that of the steam eventually obtained. A separating vessel, into which the heated water would be released through nozzles, was also required. Encountering a pressure lower than its own, a portion of the water would flash off into the steam which would drive the engine. The remainder would then be at a lower temperature through having given up part of its heat to bring about the evaporation into steam, and it would pass to the circulating pump to be forced into the coil again. A small gasoline motor (cheating a little) is used to drive the fuel pump, blower, and water pump until steam is raised.

Incredibly, this complicated reconstruction was a great success, and the McLaren engine juggernauts its way around the ground every Live Weekend.

AVIATION

Virtually all the reconstruction work for the aviation division is done by volunteer workers — road contractors, police officers, motel proprietors, electricians, decorators, students — under the direction of a permanent staff Chief Engineer. This rather unlikely collection of skills is gradually bringing the largest collection of vintage aircraft in New Zealand up to a high standard of restoration.

With 27 aircraft on hand, not to mention a hot-air balloon, numerous engines, aircraft parts, and many models, the Aviation Division requires more space than any other section. It is the aim of the museum to build up a complete collection of every type of aircraft used by the Royal New Zealand Air Force in a little over half a century of operations.

An indication of the range of the collection can be gathered from the following list of its full-size aircraft, some still under reconstruction. The dates refer to the period of service of the particular plane held by the museum.

1

S. J. Woods

1. A sketch of the flying replica of Richard Pearse's first home-made aeroplane, which had wings of bamboo covered with calico. The replica was made in 1974 for a television documentary on the life and times of Pearse.

Pearse Monoplane	N.Z. Experimental	1930's
Lockheed 10A Electra	Airliner	1936-59
Mignet Flying Flea	Ultra-light Sports	1936
Vickers Vildebeest Mk III	Patrol Bomber	1936-44
Miles Magister	Initial Trainer	1938-62
Lockheed Lodestar	Aerial Topdressing	1938-69
Ryan STM2	Initial Trainer	1938-61
De Havilland DH 89 Rapide	Airliner	1938-71
Hawker Hind	Patrol Bomber	1940-41
Curtiss P40E Kittyhawk	Fighter	1941-45
North American Harvard II	Advanced Trainer	1941-64
Lockheed RB34 Ventura	Medium Bomber	1941-47
Lockheed Hudson GRIII	Patrol Bomber	1941-46
De Havilland DH82A Tiger Moth	Initial Trainer	1942-61
De Havilland DH84 Dragon	Airliner	1943-68
Grumman TBF Avenger	Torpedo Bomber	1943-59
Consolidated PBY Catalina	Maritime Patrol	1943-53
Avro Lancaster MR7	Heavy Bomber	1945-64
Chance Vought FG1 Corsair	Fighter	1945-48
Douglas DC3	Airliner	1945-73
Short Sunderland MR5	Maritime Patrol	1946-67
De Havilland DH98 Mosquito T43	Trainer	1946-52
Miles Gemini	Communications	1948-63
Short Solent 4	Airliner	1949-60
De Havilland DH100 Vampire FB9	Fighter-bomber	1952-64
Auster 7c Antarctic	Light Transport	1955-66
Commonwealth Ceres	Agricultural	1959-69

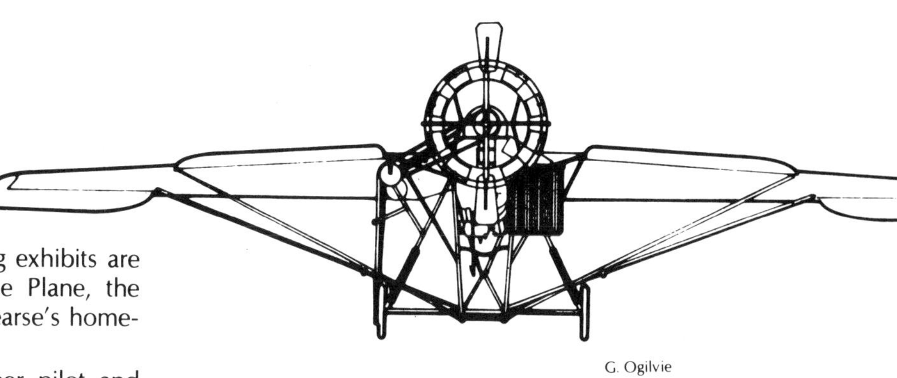

G. Ogilvie

The Pearse Planes

The oldest and most interesting exhibits are the remains of the first Pearse Plane, the complete second plane, and Pearse's home-made motorcycle.

The late George Bolt, pioneer pilot and engineer, after probing the evidence and the testimonies of eye-witnesses, concluded that Richard William Pearse made his first flight on March 3, 1904, at Waitohi near Temuka in the South Island, a few months after the Wright Brothers' famous first flight.

Fresh evidence has since been found which indicates the year 1903. It is known that after one flight, because of a heavy fall of snow that night, Pearse left his plane where it had landed on a high gorse hedge. Official records show that snow fell on July 11, 1903, and continued to fall for several days. There was no snow in that district at any time in 1904.

Thus Pearse may have flown during 1903, some months before the Wright Brothers. But whether his flights were sufficiently sustained and controlled to be accepted by historians as official flights is doubtful. Neither his early flights nor those of his contemporaries were regarded by him as being true, controlled flights. Witnesses confirm that his plane always pitched violently and always veered to the left.

On one occasion Pearse chose a friend's farm as a suitable airfield. After running a short distance, the plane lifted and, veering to the left as usual, passed over a 3-m-high fringe of willow trees and followed the course of a river before finally coming to rest on a dried-up part of the river bed. Pearse was next seen scrambling up the side of the steep river bank to rejoin the spectators.

Whatever assessment is made of his achievements, Pearse did design planes which had many advanced aerodynamic features, and with the very limited materials available to him he constructed both planes and engines and flew them.

His first aircraft was a high-wing monoplane with a wingspan of approximately 7.5 m. It was considerably modified over the two or three years it was being flown and is considered the most successful of his two aircraft. The engine was a two-cylinder, horizontally opposed type with a 10-cm bore and a very long stroke. Each

1

1. Richard Pearse in 1903. A determined and yet visionary intensity is perceptible in this studio portrait.

cylinder had a surface carburettor. The valve gear was unusual, being operated from a central point with push rods radiating out to the individual cylinders. The engine had no crankcase, the cylinders being held in position by a frame. Pearse constructed most of his own engine parts, including his own studs, bolts, and spark plugs. According to his later writings, the first engine was equivalent to a power of 18 kW and weighed about 1.3 kg per kilowatt developed.

This aeroplane appears to have been completed and tried out in 1902. It was constructed with bamboo poles which varied from 2.5- 5 cm in thickness and were available in lengths up to 3.5 m. His work was very thorough and does not appear to have suffered from structural failures. Some fittings and a small portion of the original undercarriage still exist. These were riveted and then brazed for extra strength.

The plane was supported on a tricycle undercarriage constructed of tubular steel, with the addition of three small bicycle wheels, the nose wheel being used for steering. Two posts were provided, one at the front and one at the rear of the centre wing section, from which cables were suspended to stretch to a point near each wing tip. Flying loads were taken by wire bracing running from the undercarriage outboard of the wheels to meet the upper cables at the same point on the wings.

The museum is largely responsible for the research and recognition of Richard Pearse as an inventor and aeronautical engineer. The exhibits include the remains of his first plane together with a full-size replica of the original built by Mr G. Rodliffe and Mr M. Frazer. Mr Gordon Ogilvie has had his biography of Pearse, *The Riddle of Richard Pearse*, published recently in Wellington.

In marked contrast to the other aeroplanes displayed at the museum is the fragile and ingenious fuselage of the second Pearse plane, designed and built in the late 1920s and early 1930s. This plane has been completely overhauled and reconstructed by Air New Zealand apprentices. One side is uncovered to show the interior mechanical workings.

Motat collection

1

1. Pearse's effective power cycle modifications made in 1912 or 1913 to a standard frame. No trace remains of his first major invention, patented in 1902, of a ratchet and pawl bicycle operated by a foot lever.

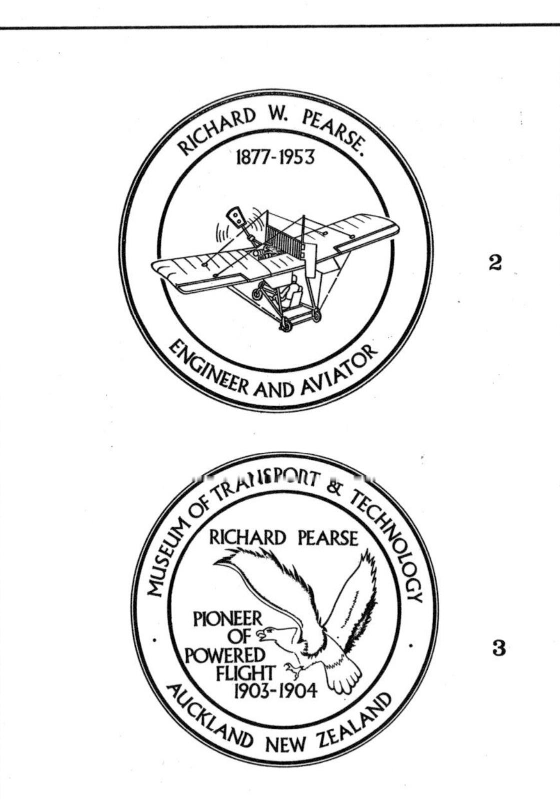

2. & 3. Two sides of the commemorative Pearse Medal struck to finance a hall to be incorporated in the new Pioneers of New Zealand Aviation Pavilion. The pavilion is to house the Pearse relics and the Walsh Memorial Library.

The Pearse Medal

One of Motat's fund-raising efforts in recent years has been the issue of a Richard Pearse Memorial Medal which has produced a substantial amount of capital for the projected construction of the Walsh Memorial Library. Pearse's aircraft are to be exhibited in this new building.

The medal depicts Pearse's first aeroplane of 1903-4, and the reverse shows an eagle and the words 'Richard Pearse, pioneer of powered flight 1903-1904'.

The Pearse medal is designed as an investment and collectors' item and is struck in .999 silver, copper, and aluminium. It can be struck in platinum or 18ct gold to special order. Members of the Numismatic Section sell the medal from special stalls during Live Weekends at the museum. The proceeds are devoted to the preservation of the Pearse exhibits.

The medal is 42 mm in diameter and suitably packed with an explanatory card. It is struck in Auckland by the Waitangi Mint and produced for the museum by the Pacific Commemorative Society.

The second aircraft, the only one that has survived in complete form, displays a remarkable number of interesting and, for the time, revolutionary features. It has a variable-pitch propeller which could be controlled manually from the cabin and is reminiscent of some hydraulic turbine vane mechanisms used early in this century. Pearse claimed the engine would develop 45 kW. It weighs only 113 kg and has two large cylinders mounted under a crankcase which houses a two-throw crankshaft. Evidently the engine was run as a two-stroke and also as a four-stroke.

Two pistons in the large upper cylinders pumped air into two smaller power cylinders. Crankcase compression fed one side of the power piston, and pump cylinder compression fed the other side. The fuel-air mixture was fired on both sides of the power pistons. The power and pump pistons were connected by a rigid rod which ran in glands at each end of the cylinders. Fuel was drip-fed to surface carburettors but a small fan blower was used to break up the fuel droplets before they impinged on the wire gauze in the surface carburettors. The engine was water-cooled, and ignition was by spark plugs, battery, buzzer, and coil. As with the first Pearse engine, lubrication was by drip-feed from numerous outlets. This engine has not run for a quarter of a century and was left in original condition when the restoration was done.

The engine is mounted on trunnions so that the angle of thrust can be varied. This and the construction of the propeller made it impractical to swing the engine by hand. Pearse arranged a belt and pulley drive from

L. Sindall

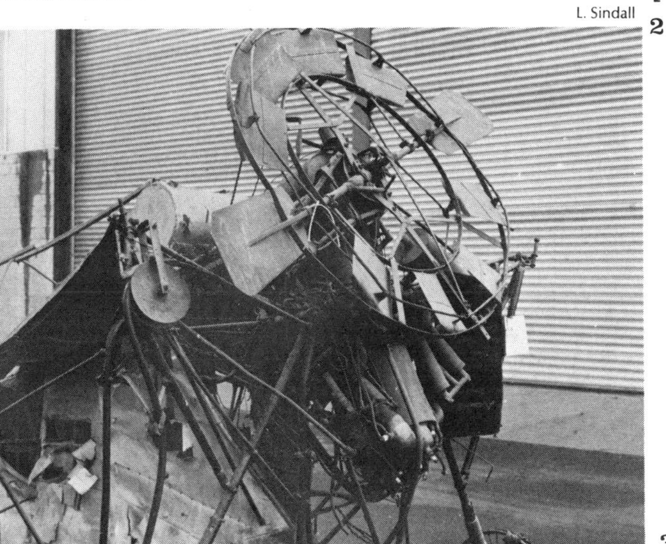

1. Pearse's Utility plane, developed in his old age, was designed for low landing speeds. It had many unusual features including a helicopter like propeller. It is pictured at Motat before careful restoration.
2. The Utility convertiplane showing the unrestored tilting and variable-pitch propeller. Engine starting was achieved by a belt drive from the road wheels.
3. A rear crankcase view of the Pearse engine that could operate in either 2- or 4-stroke mode with a 90° crank throw. Ignition wiring and priming controls can be seen. A pumping type of supercharger was fitted.

the main wheels so that by pushing the aircraft forward the engine was rotated in order to start it running.

With the engine tilted almost vertically as in a helicopter, Pearse intended that the aircraft should hover. To balance the torque when the engine was tilted vertically, a small tail rotor was incorporated. It was shaft-driven from the engine.

Leading edge flaps were intended to provide improved control while hovering, as they were located in the propeller wash area when the engine was in the hovering position. There are normal ailerons on the outboard trailing edge of the wing to provide control when it was flying with forward speed as a conventional aircraft.

The unusual construction of the rear part of the machine was apparently an expedient to simplify storage. The rear section could be folded forward on top of the cabin.

Generally speaking, this second aircraft is obviously too heavy to operate as a helicopter with the engine and propeller that Pearse has fitted. It would be a difficult machine to control and develop, but nonetheless the ideas which Pearse built into it are quite remarkable.

Motat collection

1

1. George Bolt, pioneer New Zealand aviator, at Mission Bay in 1921. He flew the first airmail (Auckland-Dargaville) December 16, 1919. He gave over 50 years' service to New Zealand aviation and was one of the founders of Motat.

2. & 3. Reconstruction volunteers working on Motat's Electra 10A. The first all-metal aircraft to fly in 1937 with Union Airways (now NAC), it carried 10 passengers at 330 km/h. Below: In the future it will be displayed as ZK-AFD, 'grande old lady' of NAC.

4. A Corsair fighter-bomber, flown by the RNZAF in the Pacific, ready for take-off on a bombing sortie. For storage the outer wing sections folded upwards at the full wing. Motat's Corsair is currently under major reconstruction.

J. Sullivan

J. Sullivan

2

Royal New Zealand Air Force

3

4

Motat collection

1. 'Left, left . . . steady.' A New Zealand bomb-aimer with a Mark XIV gyro bomb sight looks through the clear-vision panel in the nose of a Lancaster.

Lancaster Bomber

One of the greatest attractions at the museum is undoubtedly the World War II Lancaster Bomber. Because of its unforgettable shape and size and its great achievements, such as the memorable Ruhr Dams raid, it brings back memories for many ex-servicemen. Powered by four Rolls Royce Merlin engines similar to those used in Spitfires and Mosquitoes, the Lancaster played a dominant part in winning World War II.

Of the 7374 Lancasters built, fewer than a dozen remain in existence. The Lancaster Bomber at the museum was built for the Royal Air Force in England and came off the production line marked NX665 in June, 1945.

It did not see active service, however, being flown only by the French Navy. It was one of three based in New Caledonia and was flown from Noumea to Whenuapai for presentation by the French Government in 1964. This international gesture of goodwill was more internationalised by the arrival years later of extra gun turrets from Argentina and Canada.

The Lancaster flew over Northland, 'buzzing' the little school at Matakohe, before making its final landing at Whenuapai, where it was partially dismantled so that it could be moved to Motat.

1

Since then its exterior paintwork has been altered to display 1943-45 Royal Air Force war-time markings. These markings represent No. 75 (NZ) and No. 101 Squadrons, two of the fifty of this type of aircraft in which New Zealand airmen flew. The inboard engines of this magnificent aircraft are ground run three or four times each year during Live Weekends.

With all four engines working, undercarriage retraction, and the interior fitted out with contemporary equipment tended by dummies correctly dressed for 1943/44 operations, it will be the most complete functioning Lancaster left in the world.

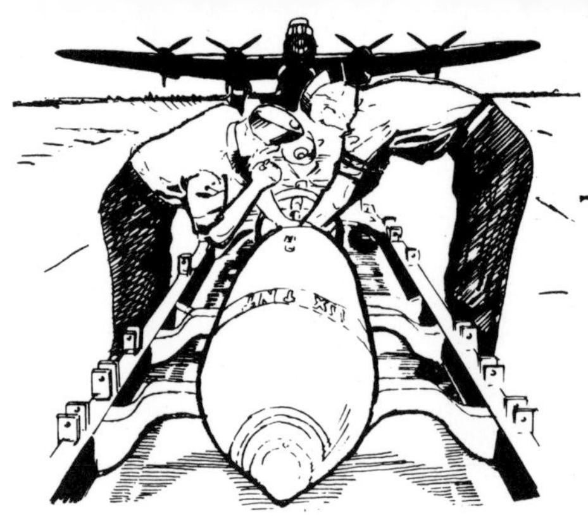

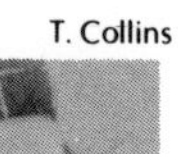

T. Collins

1. The Avro Lancaster in French Navy colours on arrival at Auckland, May, 1964, when it was presented to Motat. Two of the French crew are with Mr M. Sterling (of Motat) and his son.

2. An engine run up at Western Springs is a magnet for visitors. For better viewing and engine cooling the side cowls have been removed. The 10-m-long bomb doors are actuated during the engine running.

Opposite page:

At Motat the public can view the 5.5-m-high Lancaster from close quarters. The exterior paintwork has been officially altered using Avro's original plans to display 1943-1945 RAF markings. The engine runs during Live Weekend demonstrations.

1 2

Motat collection

Other Aeroplane Exhibits

Jet aircraft came into general use with the Royal New Zealand Air Force in the decade after World War II, and the two squadrons, No. 14 and No. 75, were equipped with the Vampire. Flying displays keep the Air Force in the public eye and are always a popular event. A breathtaking turn at displays during the 1950s was the exciting aerobatic team with their unusual twin-boom Vampires.

The aircraft displayed at Motat was one of four brought to New Zealand as ground instructional trainers in 1956 after serving with No. 14 Squadron, Royal New Zealand Air Force, in Singapore. It has a total flying time of 513 hours and 35 minutes and was acquired by the museum in 1964. The Vampire is a single-seat jet fighter bomber and was the third jet aircraft to be built in Britain. Royal Air Force Vampires established a world altitude record in 1948 by climbing to 18119 m and were the first jets to operate from an aircraft carrier.

The De Havilland Goblin engines developed a thrust of 1405 kg and a maximum speed of 870 km/h. Considering that the basic fuselage is a balsa and plywood sandwich, this in itself is outstanding.

Another of the museum's war weapons, the Sunderland flying boat, is of a type that was used very effectively by the air forces of the Commonwealth during World War II. Serving as a guardian for convoys and ditched airmen, with Coastal Command, it became a symbol of almost guaranteed safety, although some were disastrously wrecked or shot down by enemy action. NZ 4115 at the museum began its life with the Royal Air Force in 1944 and was reconditioned for the Royal New Zealand Air Force in 1953. It served in the South Pacific until 1966, with its main base at Hobsonville, Auckland.

Armed with 2270 kg of bombs or depth charges and with machine gun turrets at front, rear, and waist, the Sunderland was an effective escort and well able to defend itself. Four Pratt and Whitney radial engines, each producing 895 kW, gave the Sunderland a range of 4635 kilometres in 20 hours flying time. It is not a big aeroplane by modern standards but was a giant in its time, loved by many who flew in it or had occasion to be rescued from the sea by its crews.

Motat collection

1

2

J. Sullivan

3

Motat collection

1

M. Lane

2

Royal New Zealand Air Force

1. & 2. This de Havilland Mosquito, once flown by the RNZAF, will be rebuilt like the 'plane shown at Ohakea Air Base. The design of two Rolls Royce Merlin engines powering a plywood-balsa fuselage was a winner.

Motat's Harvard was presented by the RNZAF in 1962.

Motat collection

3. Aerial topdressing aircraft Commonwealth Ceres is refitted in the Motat workshop before display. It could carry a tonne of fertiliser or 1 275 litres of spray liquid.

4. Static engine and aircraft components are displayed in the Aviation Pavilion. An Avon jet engine is examined by the President of the Royal Aeronautical Society on a visit from England in 1974. Against the wall is a four-bladed Electra II propeller.

3

4

Selwyn Rodgers

Opposite page:

1. The 'Ops Room' evokes the World War II Bomber Command atmosphere, doubling as a show case for equipment. Shown are a radio set and the Progress Boards.
2. The twin-boomed de Havilland Vampire single-seat fighter-bomber, powered by an early jet Goblin engine, could attain 882 km/h.
3. One of the first exhibits obtained was the P40 Kittyhawk fighter, which has been restored from components collected in many New Zealand districts — the chin cowl was once a dog kennel.

Royal New Zealand Air Force

Motat collection

Motat collection

1. The long and the short of it . . . the nose of the Handley Page Hastings RNZAF transport being lowered at Ohakea. Road transport limits precluded Motat's accepting more than the nose, which was removed by saw surgery in 1972. Aircraft length is 25 m; passengers and crew 55.

2. The restored fuselage of the Pearse Utility 'plane is received at Motat in 1973. Aircraft length is 7m; passenger/crew one.

3. A yesteryear top-class airliner, the DH 89A Rapide, was brought to New Zealand in 1938 for Cook Strait Airways. ZK-AHS joined the Motat fleet in 1974. With two 150-kw Gypsy Six engines it flew 6-8 passengers at 193 km/h.

A scheduled air service across the Tasman Sea between Australia and New Zealand had long been a dream before it became reality in April, 1940. Thoughts about the service took shape soon after the famous crossing of Charles Kingsford Smith in the Southern Cross in 1928. During the years following, the Southern Cross made several crossings in both directions, as did Guy Menzies in an Avro Avian, Francis Chichester in a De Havilland Moth, and Jean Batten in a Percival Gull.

In 1935 discussions were held to inaugurate a proposed England-Australia service with a possible extension to New Zealand. In 1937 a route-proving flight was made by *Centaurus*, an Imperial Airways C Class flying boat. Subsequently three flying boats were brought out and although the war intervened the service finally began in 1940 with TEAL (Tasman Empire Airways Ltd) becoming an international byword.

As demand for the service grew, larger aircraft were required. In 1946 TEAL acquired Sandringham flying boats, which were converted Sunderlands, and in 1949 four Solents were bought.

Bigger and faster than any of its predecessors, the Solent could seat 45 passengers and cruise at over 320 km/h. It could fly the Tasman Sea in 6½ hours. The last of these custom built Solents, *Aranui* (Main Pathway), built at Belfast and carrying the marking ZK-AMO, was retained to fly the 'Coral Route' to Tahiti from 1954 to 1960. *Aranui* is now in the museum's Aircraft Park at Meola Rd.

In the Solent there are seven cabins on two decks connected by a staircase, and more space is available for cargo and mail than in earlier flying boats. There is also lounge bar and galley space. During 11 years of passenger service *Aranui* flew 4 850 000 km in 14,500 flying hours.

With her four powerful Bristol Hercules engines each producing 1490 kW, fuel consumption was approximately 160 litres per hour for each engine. This gave her a cruising speed of 400 km/h with an all-up weight of 35 800 kg.

Opposite:
Some idea of the size of the Solent can be seen in this photo taken at the museum's Aircraft Park, Meola Road.

Len

The museum's Vickers Vildebeest Mk III (NZ102) is the only aircraft of its type left anywhere in the world. Sufficient parts have been dug out of a filled-in dump to reconstruct a fuselage capable of standing on its own undercarriage. Several wingspars have been located but the ribs are damaged beyond repair. However, reconstruction is in progress and missing or deteriorated parts are being painstakingly copied and replaced. The wreckage being used comes from at least eight aircraft.

The Vildebeest was used as a bomber, torpedo bomber, and gunnery target tug. It served a useful purpose during the early days of the Royal New Zealand Air Force. Two Vildebeest squadrons of the Royal Air Force saw action against the Japanese at Singapore as late as 1941. They were powered by a Bristol Pegasus IIM3 engine of 492 kW, which gave a maximum level speed of nearly 200 km/h.

New Zealand relies a great deal on its export of meat, butter, and timber. This would not be possible if it were not for the continual feeding of the soil with artificial fertilisers. The easiest and cheapest way of spreading fertilisers in New Zealand's hill country is by top-dressing from aircraft. A number of companies in New Zealand deal exclusively with aerial spraying and dusting and over the years these firms have used an ever-improving series of aircraft designed especially for the job.

The museum displays three types of phased-out top-dressing aircraft. These are the Tiger Moth, the earliest of crop-dusters and before that the standard basic trainer for the Royal Air Force; the *Ceres*, with twice the capacity of the Tiger Moth, which was originally a basic trainer in Australia under the name Wirraway; and the Lodestar, which could carry nearly 3.5 tonnes of superphosphate and was a modified passenger liner.

The collection of aircraft at the museum steadily increases, aided by donations of obsolete or withdrawn planes from the local airlines and the Royal New Zealand Air Force. Development of the Meola Road Park progresses steadily and will receive a new boost when the tram link is established from the museum through the zoo and onto the park.

Motat collection

1

Royal New Zealand Air F

2

3

Motat collection

1. A Vickers Vildebeest patrol bomber is under construction at Motat from buried scrap parts. This 'gentle beast' with its 4.5-m propeller guarded New Zealand sea lanes in World War II.

2. Almost forgotten today are the Hawker Hinds built in England in 1935 as light bombers. By 1939 Hinds were phased out in the U.K., and 63 were reassembled in New Zealand for RNZAF to equip a flying training school. In 1941 a Hind crashed in the Tararua Ranges and in 1972 was salvaged by volunteers assisted by an RNZAF helicopter. When restored by 1978 it will be one of three left in the world, and it is hoped that the pilot of the crash will unveil it at Motat.

3. Two voluntary workers show the size of the Pratt and Whitney R-2800 Double Wasp engine on the Lockheed Ventura patrol bomber, which needs considerable restoration before being displayed.

1. The Catalina long-range flying boat has had an extraordinary life span. It was designed in 1933 and is still in use in some countries. The RNZAF operated 56 'Cats' in South Pacific units during World War II. Motat has collected major components to rebuild a Catalina.

Royal New Zealand Air Force

Walsh Memorial Library

Early one morning in February 1911, Vivian Walsh coaxed into the air his locally built Howard Wright biplane, *Manurewa I,* to become the first man in New Zealand to achieve a successfully controlled powered flight.

Together with his brother Leo, he subsequently formed the Walsh Flying School at Orakei and later at Kohimarama, Auckland. The first New Zealand pilots, many of whom served with the Royal Flying Corps in World War I, were trained here.

The Walsh brothers are recognised as two pioneers, and Motat has established a permanent memorial to them in the form of a reference library dealing primarily with aviation but also including all the forms of technology covered by the museum's collection.

The library was formed by a small band of enthusiastic members of the Aviation Historical Society, a collection of books and material being built up with donations by members and others. When the museum was established at Western Springs, the society combined its collection with the material accumulated by the various other sections to form one large collection of books, magazines, manuals, timetables, photographs, and so on.

The committee formed to run the library, comprising representatives of both the Society and museum, sorted the vast collection and under the direction of the museum librarians, Miss Lyn Mackie and Mr and Mrs Blackburn, the library is now fully catalogued under the Dewey Decimal System. Limited borrowing is permitted.

The current major project of the library is to assemble a complete collection of photographs and technical details of every aircraft on the New Zealand Civil Register, both past and present. Already some 500 photographs have been assembled. A similar project has been started on aircraft of the Royal New Zealand Air Force and on aircraft types belonging to British and foreign air forces that have visited New Zealand. Eventually it is hoped to make copies of these photographs through the Photographic Section of the Museum.

The library has many technical manuals essential to the restoration and maintenance of the valuable range of Motat's aircraft. Eight major aviation magazines of recognised authority are subscribed to, and these are bound as each volume is completed. The collection of *Flight* magazine is particularly extensive, covering nearly 70 years and complete from 1909 to 1913. A related magazine, *Engineering,* is in a series that starts with issues of 1897.

Early in 1972, Whites Aviation Ltd of Auckland contacted the Aviation Historical Society to donate the extensive collection of books, magazines, photographs, and mementos relating to the late Leo White's lifetime association with aviation in New Zealand. It was decided, in view of the extent and content of the collection, to house it as a complete unit in the Walsh Library. The collection includes an incomplete set of *Jane's All the World's Aircraft* from 1919 and a completely bound set of the magazine *White's Aviation.* The photographs range over the whole history of aviation from World War I onward, and miscellaneous items include scarce timetables, maps, and postcards. The library operation is financed through New Zealand Aeronautical Trusts Ltd and is at present housed in the upper floor of the brick pump house. Within a few years the Walsh Memorial Library will be housed in a fitting building for one of the leading specialist collections in New Zealand. This will be the three-storey Pioneers of New Zealand Aviation Pavilion.

F. Bish

2. The Lockheed Lodestar presented by Airland Ltd (now Fieldair).

HARDING STABLES
MILK
VENDOR

AGRICULTURE

Traditionally the worker on the land was a peasant of humble station in life, struggling with the crudest of equipment. Two major developments changed the pattern of agriculture and saved the increasing population of the world from famine.

One of these developments was the mechanisation of farm implements, taking the motive power away from horses and oxen and supplanting these with steam traction engines and later with internal combustion engines. The other was the escalation of scientific technology on the land and in animal and plant husbandry.

Man's inventiveness in the nineteenth century and his mastering of the techniques of metallurgy produced the first reliable machines for harvesting. Although still horse-drawn, the sickle-bar mower and combined reaper and binder, to mention only two of these advances, were a major breakthrough in releasing man from the burden of harvesting by hand.

Today, the economy of New Zealand relies on the export of primary produce. Nowhere in the world are people more conscious of modern techniques in the science and industry of farming.

Motat endeavours to display the history and technology of agriculture as it applies to New Zealand. Dairying developments are recorded by displays of butter-making churns, hand-separators, butter pats and moulds, and milk cans. Butter-making demonstrations are given on Live Weekends in conjunction with the home-made bread display in the Pioneer Village.

Tilling and earth-turning machinery is represented by a display of ploughs ranging from the earliest single-furrow type used in New Zealand to the most modern tractor-drawn model. Harvesting, always one of the more spectacular events of the farming year, is the subject of a large collection of mowers, hay sweeps, grabs, tedders, and buck rakes together with a stationary hay-bailer driven by a 1905 Hornsby hot-bulb engine.

The horses and horse-drawn equipment are the pride of the agricultural section. The horses are kept overnight at the Carrington Hospital farm and ridden to Motat by volunteers when required. At the museum they are housed in a stable building which originally

1

1: A young museum attendant giving donkey rides during the school holidays. Motat collection

Opposite page:
Spring at Motat.

stood in St Heliers, Auckland. A typical stable of a well-to-do house owner of the latter part of last century, it has room for horses, harness, and a gig or carriage, with a hay-loft above.

The horse-drawn vehicles at Motat are used regularly, principally for free rides for children. Restoration and maintenance of the vehicles and harness is a continuing exercise and one of the most time-consuming jobs of the section. Iron fittings required for wagons and farm appliances, together with the shoeing of the horses, are catered for by the smithy in the Pioneer Village.

The agricultural section maintains a collection of stationary engines of the type used to operate farm machinery ranging from milking machinery to threshing machines. Agricultural workers are also responsible for the timber and forestry section of the museum. They have built and now operate a pit-saw where planks and posts are trimmed from logs, a display supplemented by a collection of old hand- and machine-operated saws.

The variety and extent of the Agricultural Section's activities cover all phases of life in the country. Perhaps the most popular of all its shows and displays is the simplest. In the middle of the busy city, where else would a crowd of children be found but in a yard alive with chickens, lambs, pigs, calves, and other farm animals?

Motat collection

1

2

Motat collection

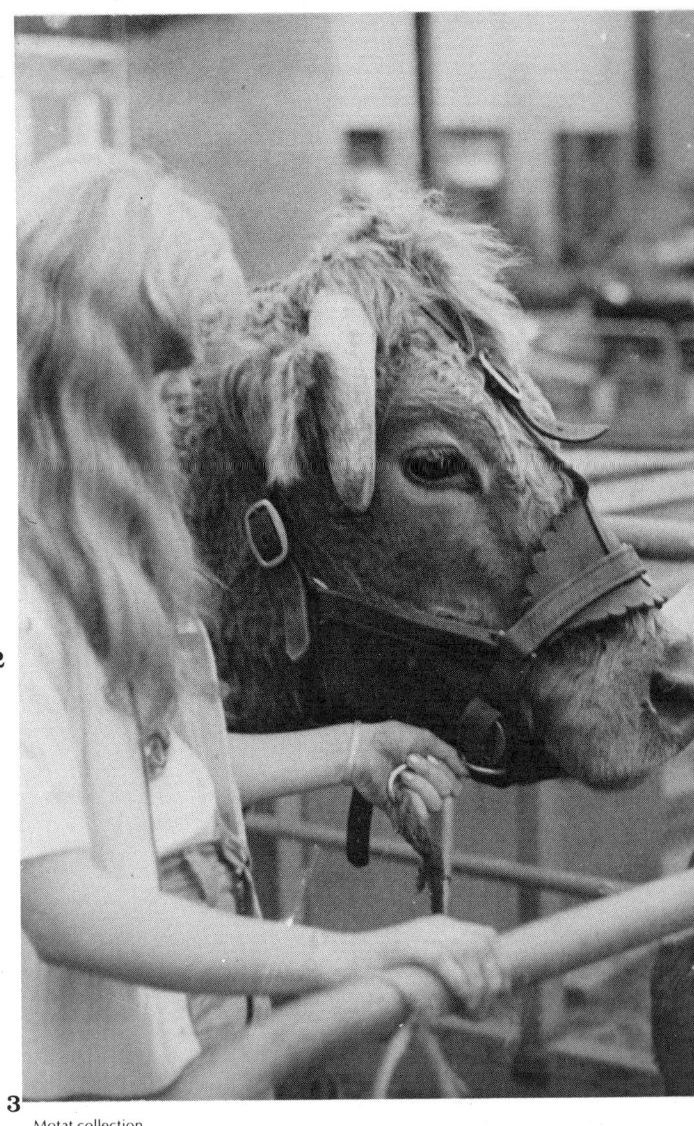

3

Motat collection

1. A demonstration of sheep-shearing during a Live Weekend, using a portable shearing plant and a wagon top as the shearing floor.

2. It sure tastes good! A demonstration of butter-making by museum members, showing methods used in the early days.

3. A 2½-year-old weighing 1 180 kg on display during a Live Weekend — part of the Agriculture Section's contribution to 'bring the country to the town'.

Tractors

The story of the development of the tractor, both wheel and crawler type, has been closely associated with the development of the internal combustion engine. The very first agricultural machines to appear were, in the main, powered by single-cylinder air- or water-cooled engines. Just after the turn of the century, tractors began to appear in large numbers and by the end of World War I were accepted as a very valuable source of mechanical power which in due course has replaced the working horse. The end of the war also saw the increase in development of many well known tractor-producing companies, some of which still market the very different-looking machines in use today.

Tractors in Motat's collection date back to the very beginning of the twentieth century. The earliest machine in the collection is a 1902 Saunderson & Gifkin two-cylinder petrol type, probably one of the first to be exported to New Zealand from England. Next comes the 1916 Samson Sievegrip made by General Motors, and then the 1919 Fordson. The latter was the first of a great series. It features the famous four-coil ignition system and flywheel so closely associated with the Model T Car.

A Cletrac Crawler dating back to 1920 is also on display, as well as a 1926 Case with a cross-mounted 13-kW 4-cylinder motor, a 1929 Massey Harris 4-wheel drive model, a 1932 John Deere with a 2-cylinder opposed engine, and a 1934 9-kW Farmall with rear-mounted mower. This last was a very popular tractor. The larger model 15-kW Farmall was also a splendid workhorse for many years.

The 1938 Fordson Model N, similar in appearance to the original model, is equipped with magneto ignition. The museum item is fitted with a rotary hoe. The 1937 Caterpillar D7 was made famous when the Minister for Public Works, the Hon. R. Semple, ran it over a wheelbarrow at a ceremony to announce the end of the old and the beginning of the new methods of earth removal and road construction.

The Fordson Model E27N was a popular tractor produced in large numbers for all purposes. The museum has two of these machines. One on steel wheels is on display, and the other on rubber is in constant use by the Aviation Section. The 1948 Bristol Crawler 20, equipped with a blade and in very old and tired condition, was persuaded to do a magnificent job levelling the rocky outcrops to make way for the development of the museum.

Among the more important tractor exhibits are also the 1958 9-kW Farmall A, which was a later model in the International Harvester range. The 1952-6 20-kW Ferguson was famous for its versatility and reliability and is still in use throughout New Zealand. The museum exhibit was fitted with snow tracks and was used in Antarctica by Sir Edmund Hillary and Peter Mulgrew, on their trek to the South Pole in 1958. It was returned to New Zealand on an Operation Deep Freeze flight in 1960.

1. Capt. J. H. Malcolm and Mr H. W. Stone upon the completion of the Sir Edmund Hillary display. The display depicts Sir Edmund and Mr Peter Mulgrew with the Ferguson tractor which they used in their dash to the South Pole, October 1957 — January 1958.

Motat collection

1

RIM SHRINKING
PLATE
WILLOW
COTTAGE
BUILT BY...
MR. WILLIAM BAGNALL
'1848'

Opposite page:

Willow Cottage as it appears in Motat's Pioneer Village. The cottage was built by Mr William Bagnall of Parnell in 1848.

PIONEER VILLAGE

Perhaps the most delightful corner of Motat is the Pioneer Village. Set in the period 1840-1890, great pains have been taken to preserve authenticity of the site and every building. With the exception of the blacksmith's forge, the original old buildings have been transported from elsewhere in Auckland to Western Springs. This has required a great deal of back-breaking labour, especially in the case of the stone cottage which was moved block by block and erected on the village site.

The village is attractively landscaped with trees and lawns on the banks of Motions Creek, the churchyard being connected to the main village area by a wooden footbridge. The village paths and roadway are paved with jarrah blocks, the last survivors of the old wooden cobblestones of early Auckland streets. A gaslight lends atmosphere to the surroundings, and a typical pioneer herb garden is tended by members of the Auckland Herb Society.

Animation on Live Weekends and other special days brings history to life. Members promenade in the village in period costume and give displays of Victorian housekeeping, cooking, and handcrafts. The authenticity of the proceedings is jealously guarded, and the food sampled by visitors is as it would have been cooked and eaten by residents of such a village in the 1880s. Beautifully ornamented loaves of bread in the shape of ears of wheat are baked by members and put on show.

In the forge a blacksmith laboriously pounds red-hot iron on the anvil, his youthful assistant manfully sweating over the large manual bellows. Meanwhile pupils in period costume toil under the stern eye of a Victorian school mistress in the one-roomed schoolhouse, learning their three Rs.

The one-room school is from Wainui, a kauri-timber district of Northland. It was built in 1878, only a year after the first Compulsory Education Act was passed.

Many pioneer children received their education in such schools. There would be only one teacher, and the children would vary in age from four to fourteen years. There was little equipment beyond a blackboard, a few maps, and possibly a small school library. Somehow the teacher would manage the numerous classes, and the children would learn to read and write and draw and do their sums.

The idea of a colonial corner as part of Motat was put forward by the Auckland Historical Society soon after the museum project was launched in 1960. The first fencible cottage was moved to the museum early in 1964.

There was some delay in finding a site for the village, but that eventually chosen includes the area where the engineer of the old waterworks lived. His house has been on the spot for over 90 years. Its two front rooms are now furnished as Victorian sitting-room and bedroom. In the latter the half-tester bed is an exhibition-piece. The trellis and vine-leaf design of the bed-head, carved out of a single piece of kauri, is a work of art and craftsmanship on the part of James Morris, who came to New Zealand in 1862 on the sailing ship *William Miles*. He spent two hours a day for two years making the bed, which is inlaid with 13,000 pieces of New Zealand wood.

There was much difficulty in finding the money for restoration, for old-time buildings were generally in very poor order when offered to the museum and had to be taken at once to evade demolition. A committee was formed of three groups interested in the work — The Auckland Historical Society, the

en Cobb

Founders Society, and the Pioneers and Descendants Club — and this has worked to acquire, repair, and furnish the present set of buildings. They do much voluntary work and raise the finance for roofing, plumbing, and building. The museum's Women's Auxiliary clean, paint, paper, and conduct visitors round the cottages and church.

The great changes in the City of Auckland which have taken place in the last fifty years — the new shopping centres, sprawling residential suburbs, and motorways — have done much to erase all trace of the pioneer settlement of the nineteenth century. The wooden structures of the early settlers have fallen easy victims to the wrecker and bulldozer, while decay and lack of preservation by owners have also taken their toll.

Because of all these changes, the pioneer village was founded with the aim of preserving and restoring a few buildings of the early days.

Willow Cottage

The most attractive pioneer building in the village is Willow Cottage, a small home which came from Akaroa Street in Parnell. It is quite a superior house for 1848 and contains most of the treasures of the village. William Bagnall was born in Rye in Sussex, England, in 1803 and trained as a ship's carpenter. He came to New Zealand in 1842 with the first Nelson settlers, moving to Auckland in 1847. When he first came north he set up himself as a builder but later turned to cabinet-making. He built this little house of well-seasoned kauri, some of it probably pitsawn. The original floors and ceiling can still be seen.

The house is very well constructed and has been altered very little over the years, except that the back porch is missing. The verandah was added about 1880. Inside are four rooms — a parlour, two bedrooms, and a comfortable kitchen-living room — all built with plenty of windows and attractive cove ceilings. Although such luxuries as taps, wash-basins, and oven do not appear, it can still be considered as a typical skilled tradesman's dwelling of the mid-nineteenth century.

Mr Bagnall and his wife had two children — Mary Elizabeth, who married John White, a renowned scholar and author of *The Ancient History of the Maori,* and Lydia Ann, who married Thomas Algar Johnson, owner of the New Windsor Hotel in Parnell. (Mr Bagnall had built the first timber hotel of this name, which was just near his house, but it was destroyed by fire and replaced by a brick structure.) Both daughters had large families and by the third generation the grandchildren numbered 44. Descendants of the Johnson family lived in the Bagnall house until 1964, when it was moved to Motat with the help of the Historic Places Trust and donations from the family.

Motat collection

1

1. Willow Cottage as it was at 3 Aotea Street, Parnell, Auckland, 1880-1890. The verandah was added later.

Potter Cottage

The old stone Potter Cottage has been a landmark in Auckland for over a century, standing on the Onehunga side of the Greenlane-Manukau Road intersection. The land here had been bought by George Graham of the Royal Engineers, Superintendent of Works under Governor Hobson. He paid £131 for the three hectares and two years later sold it to William Potter, a successful farmer with a large holding in Epsom, including the present site of Alexandra Park Raceway, then known as Potter's Paddock.

William Potter came from Durham and had sailed his own schooner to the Bay of Islands in 1833, whence he moved to Auckland. He grew a great deal of wheat and grazed cattle, and as early as 1848 a Livestock Show was held on his land. In the same year, some of the military pensioners were given temporary accommodation in his barns while their cottages were being built.

William Potter had his Epsom cottage built in 1850, but it does not appear that he lived in it himself. It is very small, although it originally had a large basement kitchen. It was rented to Joseph Bycroft, the millwright who from 1855 to 1859 helped his brother to build the flour mill in Mt Eden. After this Mabel Potter, who became Mrs Lennard, lived in it until 1875.

Only the front of this house was built of squared stones of volcanic rock, the sides being rubble held together with a lime mortar. All the walls are very thick. It is not clear whether the solid construction of this cottage was adopted with a view to possible defence needs or just because the volcanic stone was so plentiful in the area. Certainly comfort was not considered greatly, as the basement kitchen was very dark and the sleeping accommodation limited.

When it was removed from Manukau Road in 1969 to allow for road widening, Potter Cottage was still owned by Potter's descendants. It is not often that an Auckland house remains so long with the same name on the title deed. The Potter family contributed towards the cost of its removal to Motat and towards its preservation.

There are a few other stone houses left in Auckland, all very well built, and the high wall of the old Albert Barracks as well as the Mt Eden Prison were constructed of these long-lasting blocks.

The Smithy

Every large settlement in New Zealand during the pioneer period had at least one smithy. The smithy would shoe the horses and was expected to repair the ironware on carts, other forms of transport, farm implements, and household items. He worked with the wheelwright and flour miller and often with the engineers in the early gold mines and batteries.

The forge or smithy was always the centre of attraction for the small boys of the village, and the adults also found it a warm and pleasant place for a little gossip on the affairs of the neighbourhood. Since forges are usually too old or derelict to be transported, this

1

N.Z. Herald

2

Motat collection

1. The blacksmith, the late Mr Fred Pasley, demonstrates the methods used to make horseshoes in the museum forge.

2. The Village forge, equipped as a working exhibit and used extensively during weekends and holidays.

forge was built largely of new material, but the tools and bellows came from an old nineteenth-century forge.

Fencible Cottages

The fencibles were pensioners from various regiments of the British Army, all of whom had seen over 20 years of active service in India, China, Afghanistan, or other countries. Although no longer fit for the rigours of overseas campaigns, they were all under 48 years of age, of good physique and character, and quite fit for the garrison duty which was theirs when they came to New Zealand in the years 1847-8, a few following in 1852.

The state of affairs in this country in 1845 was extremely precarious. Some Bay of Islands tribes, led by Hone Heke, were in rebellion and had destroyed the town of Kororareka, the former capital. The Governor felt unable to protect settlers since he had only 1100 trained troops in the whole country. He immediately asked the Colonial Office for 2000 more soldiers. This would have meant a great deal of added expense to an already overburdened colony, so a plan to use recently retired soldiers was put forward as an alternative. A new regiment to be called the Royal New Zealand Fencibles was to be formed during 1846, and the conditions of enlistment were posted in every camp in Britain.

Two of the cottages provided for these settler-soldiers are at Motat. Both were built in 1848. Sergeant Quinlan's cottage is a single-unit house with a large attic and was brought from Ireland Road, Panmure. Quinlan was the Fencibles' Paymaster who enlisted in Maryborough in Ireland. He had served with the 49th Foot and arrived in New Zealand in the *Clifton*. His family of six lived in the cottage.

The other cottage is a double unit, one half of which belonged to Private Daniel Lawlor, an Irishman who had served in the 95th Foot and who also came in the *Clifton*. The cottage was erected in Lagoon Drive, Panmure, and by the time it was acquired by Motat only one half remained. Old kauri timber was used to reconstruct the missing section, and it was made into a domestic museum displaying the utensils used for cooking, washing, ironing, butter-making, dress-making, and so on.

1

Motat collection

1. And this is how the cottages arrived. This picture shows Quinlan's Fencible Cottage being unloaded from a transporter upon its arrival from Panmure.

Opposite page:

Pictured at the porch entrance of the Chapel of the Good Shepherd are the Pioneer Village schoolchildren, who take part in Live Weekend activities in the Village.

S. J. Woods

S. J. Woods

Motat collection

The Village Church

In 1974 the Right Rev. E. A. Gowing, Bishop of Auckland, re-opened the Chapel of the Good Shepherd at Motat. Formerly St Saviour's Anglican Church at Blockhouse Bay, the church is preserved at the museum as a typical example of the places of worship used by pioneer congregations.

The Chapel of the Good Shepherd has an interesting history. In 1867 Bishop Selwyn set aside a site in the district of Waikomiti at what was then known as Whau Bay. The site was not far from a blockhouse built to protect settlers during the Maori Wars. Many years later the district became known as Blockhouse Bay.

It was not until 1898 that the church was finally completed and dedicated. It was built by Mr J. A. Penman (a baptist) who was assisted by Mr John Herbert (a methodist). The first marriage took place in the chapel in 1898. The chapel in its original situation at Blockhouse Bay did not have a graveyard in keeping with other churches of the period. However, a replica churchyard of the time has been created at the museum and actual pioneer gravestones have been placed near the church.

The altar was given to the old church by the Rev. H. R. Jacks in 1923. The pews came from old St Mary's Cathedral in Parnell and still bear numbers and nameplates, indicating the old system of paying pew rents which was in vogue until the early 1940s. The fine pair of sanctuary chairs are the gift of Mrs Winifred MacDonald. The eagle lecturn, carved from solid English oak, was originally in St Thomas' Church, Freeman's Bay. It is on loan to Motat from St Matthews-in-the-City. The candle sticks and missal stand are from old St Martin's, Mt Roskill. The missal is from St Luke's, Mt Albert, and the Prayer Book from St Jude's, Avondale. The font base, Credence, hymn board, candle-snuffer, and brass altar vases are from St Luke's, Parakai. The brass alms dish is from the old St George's, Kingsland, and various embroidered items plus the bookmarkers were given by St Albans' of Balmoral.

Another interesting feature of the church is the display of house flags of shipping lines which plied the Manukau Harbour. They are hung there in memory of the sailors who served on the West Coast of New Zealand. The lines represented are the Holm Shipping Company, Anchor Line, Northern Shipping Company, and Union Steamship Company.

Near the entrance to the church is a display area with a collection of historic bibles, prayer books, certificates, and Sunday school material.

During Live Weekends, organ and choir recitals are presented in the church

Further developments of the Pioneer Village are anticipated. The Cropper House, a fine, superior Victorian residence, has been acquired, and long-term plans have been formulated for the inclusion of a hotel and other buildings in the village.

Without a doubt the Pioneer Village will remain one of the greatest attractions of the museum. Preserving the story of the life of our pioneer forefathers is part of the conscience of the nation.

N.Z. Herald

1. Museum staff preparing old headstones for placing in the graveyard beside the Chapel of the Good Shepherd.

Ladies Auxiliary

Since the beginnings of Motat, there has been a large voluntary involvement in practically every section by girls and women of varying ages and walks of life. Visitors are amazed to see overalled lasses covered in grease and grime working alongside the lads on most unpleasant cleaning jobs, like that of preparing a veteran locomotive for repainting. More amazing still is the transformation of enthusiastic toilers into lovely ambassadors of the female sex on museum social occasions.

Women at Motat are aware of the failing of most men in that they seldom do the little things that really matter. The women run their practised eye over Motat and direct their attentions to good housekeeping. In particular, the cleanliness of the Pioneer Village is a credit to the Ladies' auxiliary, who spring-clean, scrub and polish every day, and tend to the gardens regularly. The cottages are as clean and tidy as if they were actually lived in, and on Live Weekends and other special occasions the women do live and cook in the pioneer homes. Most of the ladies have painstakingly made their own costumes and wear them while tending the village. This creates a lovely atmosphere of pioneering days — a walk through the village is a journey into the last century.

Elsewhere in the museum, ladies are involved in the Tramway Section, both in restoration work and as conductors. To date there have been no licensed lady tram-drivers, but there has been a lady bus-driver. Young girls frequently conduct on the double-decker buses. Members of the fair sex are also heartily involved in the Railway, Aviation, Numismatic, Printing, and Agriculture Sections.

A husband-and-wife team spent nearly two years restoring a 1902 Saunderson & Gifkin tractor, possibly the oldest farm vehicle in New Zealand. Another husband-and-wife team has developed the Walsh Library from an unrelated pile of books into a well catalogued reference library of note.

It would not be possible to run the museum on weekends and other holidays without the devoted assistance of ladies in the office, souvenir shop, and cafeteria. Perhaps the hardest working of all are the few ladies who 'feed the troops' during Live Weekends. These dedicated ladies work under very difficult conditions and are besieged three times a day by ravenous members.

The Ladies Auxilary committee consists of all the ladies of Motat who are summoned together by telephone on special occasions. A meeting is seldom convened, but members continually meet in association with their various duties and interests at the museum.

Motat collection

2. A wheaten loaf made in the Quinlan Cottage during a Live Weekend demonstration.

1

Motat collection

2

Motat collection

3

Motat collection

Motat collection

4

1. Mrs Joyce Lush and her daughter Barbara, dressed in period costume and photographed on the verandah of Willow Cottage.
2. A museum member in period costume shows her small son just how the old village pump works.
3. All in keeping with the period — Mrs Barbara Stevens promenades with her baby, Timothy.
4. Motat's souvenir shop is a popular place for buying gifts, many of which have a distinctive museum association.

Shopping Street

In its formative years Motat amassed a large collection of exhibits which fell roughly into three catagories — the Victorian era, pre World War 1, and pre World War II. The two wars ended one life style and heralded a new, the changes rung being greater than in any similar period in history and the modernisation practically world wide.

To display Motat's collection in modern glass cabinets would have been to lose the value of the exhibit material, so a policy was adopted to display wherever possible in the correct period environment. The Victoriana was faithfully displayed in the Pioneer Village cottages, thus preserving the atmosphere of the period. For the later material it was decided to build a street typical of the period from the turn of the century to the late 1920s.

Motat collection 1

A great deal of research went into the planning of the street, authenticity being obtained from early photographs of the shopping areas of Auckland. Clever profile drawings were used in lieu of plans and working drawings, and a builder with a genius for improvisation constructed the first three shops from material from demolition yards, blending it with authentic stables which were shifted in sections from St. Heliers. The site chosen for the shopping street was by the tram tracks; it is appropriate to have the electric tram cars rumbling by just as they did so many years ago.

The first shops were a chemist's, a general store, and a drapery, with a ladies' and gentleman's hairdressing parlour upstairs. The shops were ficticiously named after prominent museum members at the time of construction in 1972. They were painted with advertisements of the period.

Motat was fortunate in purchasing a great many chemists' jars from one of the earliest businesses in Auckland, which, added to the material already held, produced a fine initial display which is steadily being increased.

2 Motat collection

1. An artist's conception of the turn-of-the century shopping street in the museum.

2. Going shopping.

Opposite page:

1. The engraver at his trade in Motat's jeweller's shop, which was patterned on the Queen Street shop of James Pascoe, 1906.

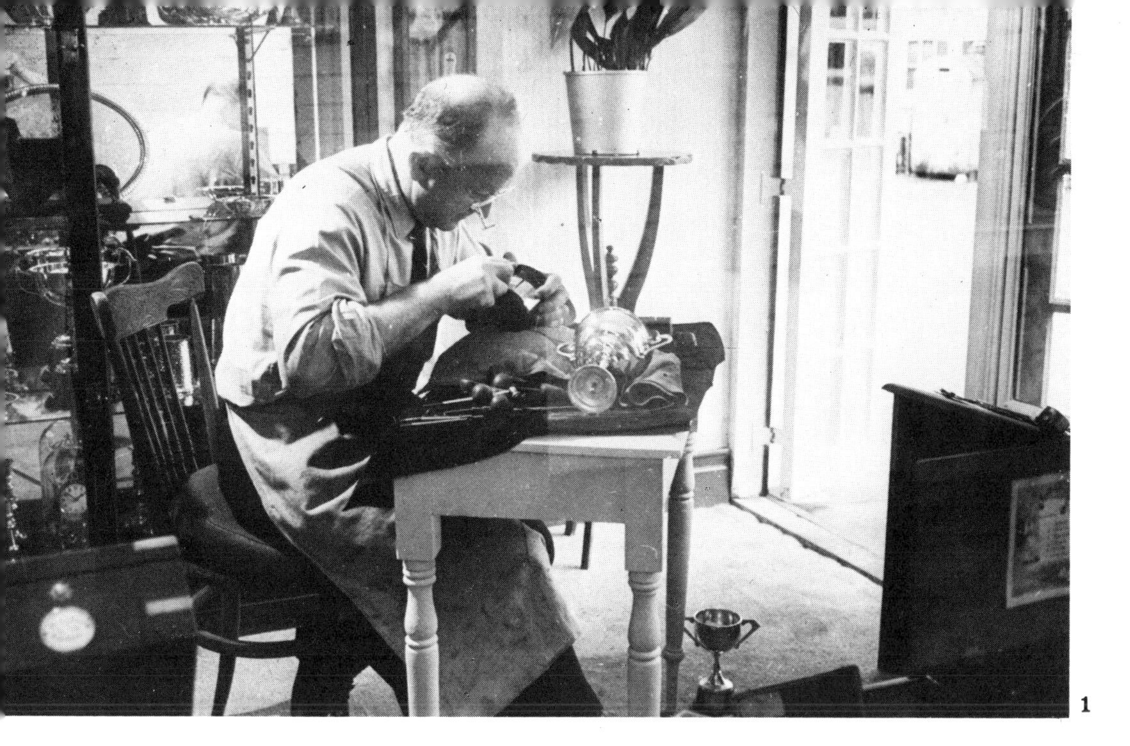

1

The general store, in those days, sold everything. Here you will see brand names long since forgotten. Nearly a 100 different brands of tobacco and cigarettes and 23 brands of butter were manufactured in Auckland alone. Presiding over the general store is a spring-loaded money conveyor.

The draper's shop is a treasure chest of haberdashery from ostrich feathers to cottons and silks and other women's apparel long since outdated.

Up in the barber's shop, where a shave cost ninepence and a haircut sixpence, the barber is talking to a customer in the ornate hydraulic chair. It is interesting to note the price of ladies' bobs — sixpence — and shingles — ninepence.

The shops are faithfully tended by women members who clean and change the displays regularly. The window display of the chemist's shop may one week feature laxatives — the senna pods and enemas and dreadful blue bottle of castor oil — and the next week babies' food and feeding paraphenalia. The collection is constantly being added to by well-wishers, who are really delving into the past on back shelves of old shops and into long forgotten cupboards. Every now and again a real prize is discovered — a precious packet of St. Mungos Soap or a rare box of wax Vesta matches.

Stage two of the street was completed in 1973 with the addition of three buildings sponsored by the actual firms who established them a century ago — Pascoes jewellery, the Bank of New Zealand, and Hellaby's Butchery. While collecting items for the butcher's shop a 1913 price list was discovered. This has been reproduced and is displayed with joints of meat and smallgoods in the window — cheapest sausages in town at sixpence per pound and rabbits hanging on the rail at ninepence each.

Linking the two blocks is a replica motorbus depot of the 1920 period. It is named to commemorate two famous rival passenger transport firms of the period — Aard Motor Services and White Star Motors. Parked in the depot is a genuine 1923 North Shore bus.

Not far away is a solicitors' office, authentically furnished as a turn-of-the-century exhibit. Here, two early carbon electric light bulbs struggle to illuminate the dingy premises. Elswhere in the street there is a balance of gaslight and early electric light.

2. Where does this go? Mrs Margaret Hutchinson arranging the display in the newly built replica chemist's shop.

1

1. A Live Weekend display demonstrates the old deep-sea diving suit.

1

Motat collection

2

Motat collection

3

N.Z. Herald

4

Motat collection

1. The Wainui School in the museum holds regular classes during Live Weekends, using teaching methods of the early 1900s.
2. A happy group of young visitors having a free ride on the horse-drawn wagon during a Live Weekend.
3. Oops, it's a long way down! Linda Rex tries her hand at riding the penny farthing bicycle.
4. The old and the new in engine tune-ups. Motor Specialties Ltd presented Motat with the first engine combustion machine to be used in New Zealand.

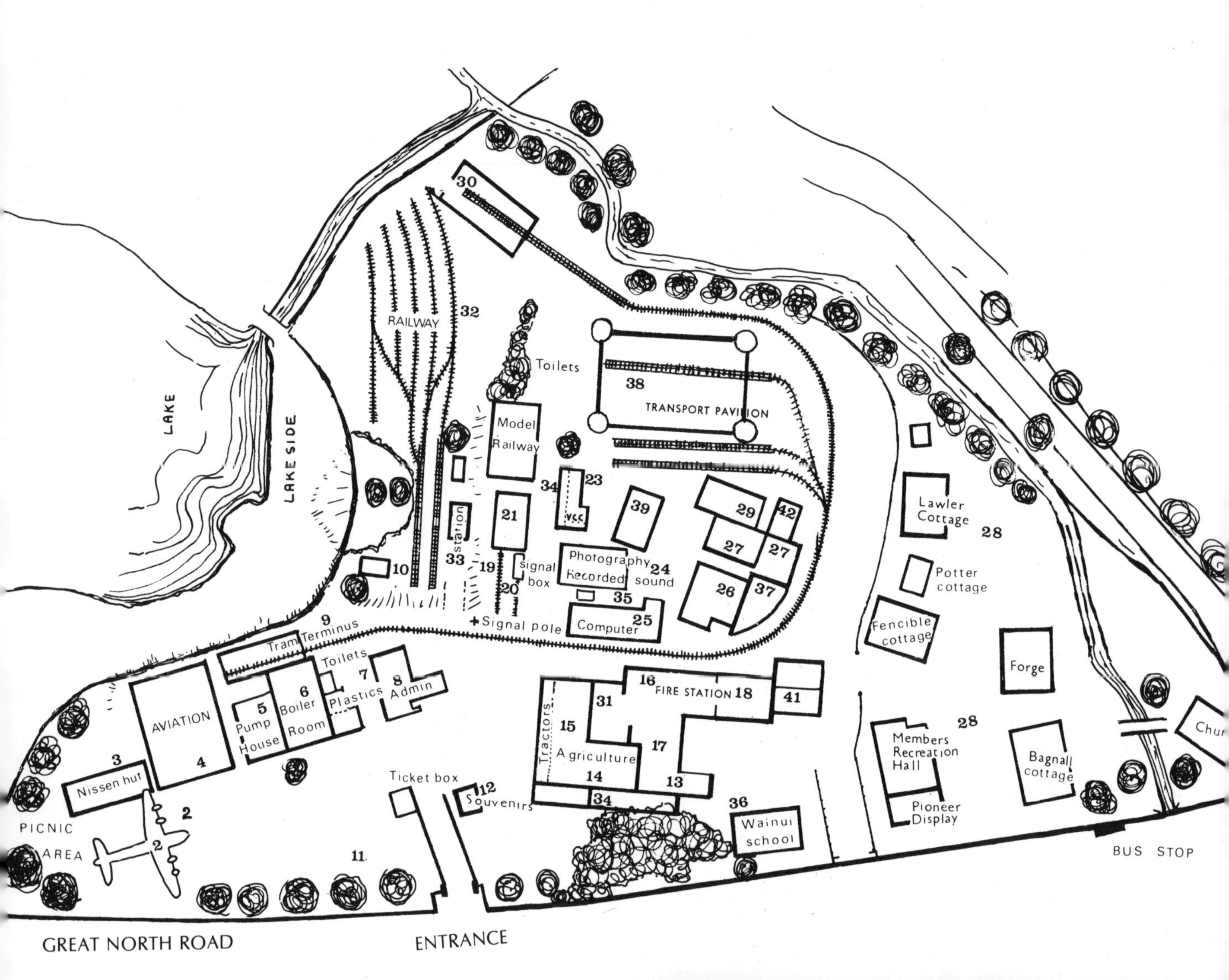

LAKE
LAKE SIDE
RAILWAY
32
30
Toilets
38
TRANSPORT PAVILION
Model Railway
34
23
VCC
39
29
42
27
27
21
station
33
19
signal box
20
10
Photography
Recorded sound
24
35
Computer
25
26
37
+ Signal pole
9
Tram Terminus
Toilets
7
8
Admin
Plastics
6
Boiler Room
5
Pump House
AVIATION
4
3
Nissen hut
2
2
PICNIC AREA
11.
Ticket box
12
Souvenirs
16
FIRE STATION
18
41
31
15
Tractors
Agriculture
17
14
13
34
36
Wainui school
Lawler Cottage
28
Potter cottage
Fencible cottage
Forge
28
Members Recreation Hall
Pioneer Display
Bagnall cottage
Chur
BUS STOP
GREAT NORTH ROAD
ENTRANCE

MOTAT

Legend

1. PICNIC AREA
2. LANCASTER BOMBER
3. "NISSEN" HUT, AVIATION EXHIBITS
4. AVIATION PAVILION
5. BEAM ENGINE ROOM, INCLUDING STEAM AND WATER SUPPLY EXHIBITS
   Upstairs—BUSINESS EQUIPMENT, GRAMOPHONES, DOMESTIC AND LIBRARY
6. STEAM ENGINEERING DISPLAYS
7. PLASTICS INDUSTRY
8. ADMINISTRATION AND CAFETERIA
9. TRAM TERMINUS AND WORKSHOP
10. NATIVE SHRUBS ETC. BUSH RAILWAY LINE
11. RADAR SCANNER AND GUNS
12. PLAYGROUND
13. AIRCRAFT ENGINE WORKSHOP
14. AGRICULTURAL IMPLEMENTS AND MACHINES
15. COMMUNICATIONS AND ELECTRONICS
16. FIRE STATION
17. AVIATION WORKSHOPS AND STORES
18. MECHANICAL WORKSHOPS
19. LOCOMOTIVES
20. NZR SIGNAL BOX
21. RAILWAY DISPLAY
22. MODEL RAILWAY CLUB
23. VETERAN AND VINTAGE CAR CLUB BUILDINGS
24. *Upstairs* — PHOTOGRAPHIC AND RECORDED SOUND
    Downstairs — PRINTING DISPLAY
25. HALL, COMPUTERS, SCALES, CLOCKS, HOUSEHOLD AND OFFICE EQUIPMENT
26. STAFF WORKSHOPS
27. STORAGE AND STABLES
28. PIONEER VILLAGE AREA
29. TRAMWAY WORKSHOP AREA
30. LOWER TRAMWAY TERMINUS
31. SPECIAL DISPLAYS
32. RAILWAY TRACKS AND EXHIBITS
33. RAILWAY STATION
34. STATIONARY FARM ENGINES
35. ARMY HUT
36. WAINUI SCHOOL BUILDING
37. SHOPPING COMPLEX (1900-1920)
38. TRANSPORT PAVILION
39. ARMAMENT
40. DUTCH ORGAN
41. CELL BLOCK
42. TRACTION ENGINE

B. O'CALLAGHAN
HAIRDRESSER
J.H. MA
GENERAL
CREAMOATA
THE NATIONAL BREAKFAST
DRAPE
WAXS

COMMUNICATIONS

A long mural depicting the development of the telephone from the earliest types of semaphore to world wide communication via satellite and modern, instant transmissions is the main feature of the pavilion devoted to telephony, telegraphy, radio, and television. The mural is some 15 metres in length and complements displays in the five exhibition bays beneath. Each bay shows examples of communication equipment of ever increasing sophistication and complexity.

Starting with the purely visual communication of the semaphore and the heliograph before 1840, various steps are taken in the development of the electric telegraph and the code invented by Samuel Morse. This was the very birth of data transmission and was developed by fine Victorian engineers who built their equipment in the tradition of the bygone clockmaker, producing many functional and beautiful instruments.

The progression of pictures and exhibits moves on to Alexander Bell's audio experiments for the deaf and to the first telephones of the 1870s, made to the same principle design as those used today. By 1900 morse and telegraphy had declined in favour of the telephone which was then an elegant instrument with unrealised social effects. It was just beginning to gain popularity and use in the home.

With this period of rapid expansion, skilled craftsmen who produced instruments of elegance and precision began to give way to greater production demands, and the more functional and less ornate telephones become evident. Ugliness became commonplace as automatic dialling became more important than shiny brasswork. Telephones on display cover the whole range of instruments used in New Zealand starting with the Ericsson open table model of the 1880s. This has the earliest type saw-tooth centrifugal generator cut-out.

The next great advances are the high-speed morse system of Murray and Creed, the first radio transmitters and receivers, the gearing for better communications in wartime. Early radio equipment includes a Marconi Multiple Tuner of 1912 (patented 1907). This is a ship's receiver used in conjunction with a magnetic or carborundum detector.

Centrepiece in the radio display is a small complete radio-transmitting station used in New Zealand in the early 1930s.

An extensive display of radio valves includes some rarities, including a Fleming type diode made by American Marconi in about 1912. This was invented by J. A. Fleming in 1902 and was the first radio valve. The de Forest Spherical Audion of 1907 was invented the year before by Lee de Forest in the United States. It could perform the three basic funtions of amplification, detection, and oscillation.

There are several examples of the original Audiotron, the first valve to be made available for sale to amateurs and experimenters. It was made by E. T. Cunningham of San Francisco in 1919. The famous R valve is shown in its various stages of development from the French military wireless type of 1916-18.

A forest of overhead wires began to change the environment as man entered an era of vast change and development. Depicting the great developments which took place between the two World Wars, displays include the telex, now an essential part of big business and news reporting, the first transmission of radio pictures, and the PABX system. This was the heyday of the central battery system and the end of the morse era.

The final section of the mural shows the post war period with its beginnings of profound change. In this period are the videophone, memory phone, laser, glass cable, and the global transmission of data, television, and telephone. And so into the future.

The museum exhibits cover early telegraph and telex machines, telephones, radio generators, cables, televisions, valves and all the impedimenta of communication in an electronic world. Perhaps the most interesting of the telegraphs is the Murray Multiolex Machine Telegraph System for keyboard transmission of telegraphic messages. This was once extensively used throughout the world and was the brainchild of a New Zealander, Donald Murray.

In the same category is the Creed Keyboard Perforator, used for the preparation of perforated tape with holes representing dots and dashes. When fed into an automatic transmitter, the perforated intelligence was converted into electrical impulses for transmission.

n Guard — a Motat member dressed in the
iform of a bandsman of the Wellington Rifle
lunteers, 1880, stands guard at the Village gate.

1. Posters produced on period presses by volunteers at Motat.

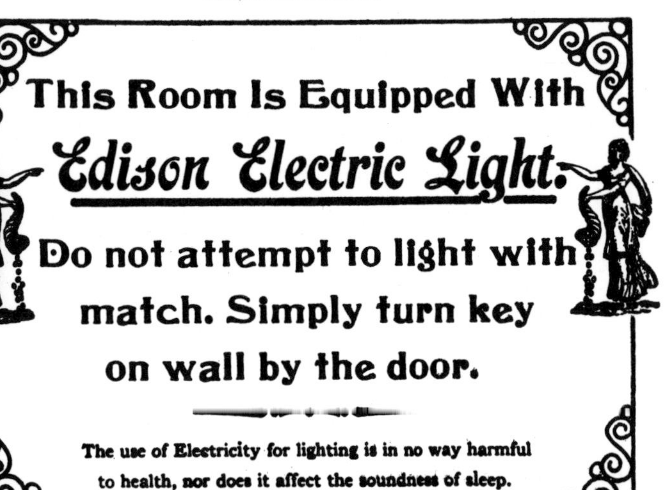

1

Printing

One of the first industrial groups to support Motat was the Auckland Master Printers' Association and the Auckland Guild of Printing House Craftsmen, who sent representatives to the initial meetings called to set up Motat.

The Printing Section was set up in 1964. While the bulk of members of the section were from the Guild, the Master Printers' Association was the real administrative force. In 1965 the Printing Section became a sub-committee of the newly formed New Zealand Institute of Printing.

The first piece of equipment acquired by Motat was an 1896 Alexandra toggle-platen press, from the School of Printing in Auckland. The original workshop and store of the section was the front room of Willow Cottage in the Pioneer Village, a rather cramped area shared with old tram seats, a bath, and other items.

Soon a stone lithography press of the 1920s was offered to the museum. It is a vast piece of machinery weighing several tonnes and too big to be housed under prevailing conditions at the museum. It is still in storage at the Naval Base, Devonport, awaiting the day when it can be transferred to Western Springs.

Motat's most important printing acquisition has been the entire plant of the Hokianga Herald, Kohukohu, Northland. This country newspaper was established in 1906 and ceased to function in 1953. A group of printers from the section went to Kohukohu and dismantled the machinery, transported the pieces 40 km to the nearest railhead, and brought the entire workshop to Auckland. The last forme to be printed was discovered still sitting on the bed of the machine. The last owners had just walked out and left everything as it was.

A Wharfedale, flatbed, stop-cylinder press was the next to arrive for the collection. This was followed by a cropper plateau, a hand-operated guillotine, and type cabinets complete with cases and type. It is planned to use this equipment to recreate the Hokianga Herald workshop.

A Babcock cylinder press was found in Cambridge, where it had been used as a newspaper press. This is also in storage. A Heidelberg Plateau press of 1925, the first to be imported into New Zealand and one of the oldest of its type in the world, is under restoration.

1. A member in action on the vintage printing presses.

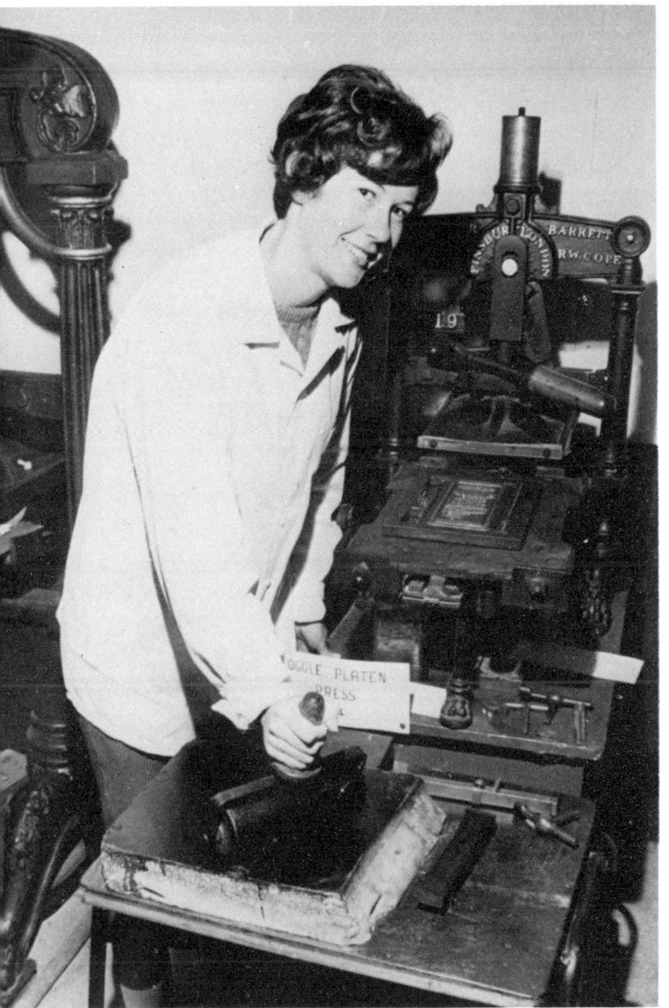

1

Motat collection

The following items consitute the main section of Motat's printing display:

Hopkinson's Improved Albion Press, 1834. Made by Jonathan and Jeremiah Barrett of Finsbury, London, this type of press was to become the most popular of hand machines, surviving to modern times as a proof press and manufactured as late as 1940 by Dawson, Payne and Lockett.

Alexandra Printing Press, 1876. Donated by the School of Printing, Auckland Technical Institute, this press was named after Princess (later Queen) Alexandra. The toggle-action machine has a bar and knuckle joint which are used to exert pressure on the platen or top plate. When the pressure is released, the platen is returned to its original position by means of a spring which is housed in a cap in the form of Prince of Wales feathers.

Model Platen Press, c.1900. This type of press was first displayed at the Caxton Exhibition in 1877 by C. G. Squintani & Co. of London. The motivation is by treadle, this type of machine being used mainly by job printers, amateurs, schools, and missionaries in remote areas.

Wharfedale Stick-fly, Stop-cylinder Letterpress Machine, c.1900. Made by Payne & Sons, this was donated by the New Zealand Government Printing Office. The cylinder comes to rest after each impression to allow the carriage to return for inking.

Chandler Price Platen, c.1900. Donated by Clark & Matheson Ltd, this is a card and billhead press with a vertical bed and platen, the latter being hinged below the lower edge and arranged to rock back and forth.

Linograph Typesetting Machine, c.1908. This was donated by V. Parmenteir and restored by C. Lewis of the School of Printing.

Shaw Honley Pen Ruling Machine, c.1918. This machine was used for ruling account books, ledgers, and similar works requiring ruled lines, both horizontal and vertical. Mechanical ruling dates from 1770.

Heidelberg Platen Machine, 1925. On loan from A. M. Satterthwaite Ltd this machine, 25 x 38 cm, has an exceptionally low serial number (338). Internationally, the machine is extremely rare in fully operational condition such as this.

Harris Offset Lithographic Printing Machine, 1945. This item is on permanent loan from Whitcoulls Ltd. It is an early example of a modern sheet-fed offset lithographic printing press.

1

1. Sections of the IBM 650 Data Processing System — the first electronic digital computer used in New Zealand (1953) in the N.Z. Treasury, Wellington.

Barry McKay

Computers

Motat has several unusual and historic computers and calculators on display in the Computer Hall. The oldest of these is an automatic totalisator dating from 1913 and the largest the IBM 650 Data Processing System.

The totalisator is a Julius Premier invented by Mr (later Sir George) Julius. The Auckland Racing Club used this machine in March, 1913, thus gaining the distinction of being the first in the world to use a fully automatic totalisator betting machine. The invention arose from experiments to devise a system for automatically recording votes at parliamentary elections, and from the very first day of operation it was a great success.

Perhaps the machine with the most visual impact is the Meccano Differential Analyser, built at Cambridge University, England, in 1934. This incredible device has been made entirely from Meccano parts favoured by boys for generations as an educational toy. Designed before the advent of today's electronic digital devices, it is an analogue type measuring computer which can provide answers to what were then otherwise unsolvable problems. It was used in Cambridge until 1939. Similar machines were used extensively during World War II on operational research problems including the design of bombs used by the Dam Busters.

Motat's machine was brought to New Zealand in 1950. It was first used by the Seagrove Radio Research Station and later by the Dominion Physical Laboratory for problems associated with hydroelectric power station design.

The IBM 650 digital computer was one of the first production models, announced in 1953. The machine on display at Motat was installed at the New Zealand Treasury in 1960 and comprised the 650 console unit (central processing unit), 727 magnetic tape units, 533 card read punch, 407 punched card accounting machine, storage, power and control units, and a sterling conversion unit. The main components are on display in the museum.

Photography

The fact that photography is not by any means a modern hobby or skill is shown by the great range of cameras and photographic equipment, covering more than a century, in Motat's display. Early examples of negatives and prints on display show that in the field of general photography our grandfathers could do just as well as the modern enthusiast with his intricate and expensive gadgets.

The first camera of a practical kind was produced by an Austrian, Voigtlander. This consisted of a long brass tube, wider at one end, resting on a stand. The tube accommodated the sensitised plate through a slot in the body. Even today, lens manufacturers use the calculations of Josef Petzval (1807-1891), a Hungarian mathematician whose lenses were used from the very first.

The original sensitised plates were of a very highly polished silver and became 'positive' by the action of mercury vapour. This system was invented by Louis Daguerre in 1837 and hence photographs were known as Daguerreotypes. Motat's display includes a number of Daguerrotype portraits of mid-Victorian men and women.

Most of the cameras on display are half-plate or whole-plate size. This indicates that the size of the photograph or negative produced is 16.5 x 12 cm or 21.5 x 15 cm. These are mostly made of wood, brass-bound in places, and are in themselves fine examples of the cabinetmaker's art. Three of these large cameras are displayed together on their tripods to show the essential differences between the English, German, and American types of the same period.

The English model looks rather fragile and is equipped with the well tried and almost infallible Thornton Pickard roller-blind shutter — a sprung, opaque blind with an opening in the centre. As the spring is released, the opening passes across the lens to expose the plate. Speed times are regulated by various spring tensions. The German type has a built-in metal 'leaf' shutter and the American an early type of between-lens shutter, the speeds in the latter being controlled by an air piston which delays or speeds the shutter by a valve control. The slides of all three cameras were made to receive glass plates.

Another exhibit is a Goertz Press Camera using plates 21.5 x 16.5 cm. Both large and small models of this camera were used by the press throughout the world, but the cumbersome size of the large model must have severely restricted its use.

The Thornton Pickard triple-extension, bellows, half-plate camera is a good example of an early device for close-up photography. It was used for photographing postage stamps, medals, and jewellery.

The name Kodak is synonymous with photography. George Eastman (1854-1932) saw the possibility of making photography an easy exercise for the man in the street. He devised a simple camera with a special lens and foolproof shutter that a person with no prior knowledge could master with success. The cameras were supplied with a roll of film inside, and the user later returned the camera to a depot where the roll was taken out and replaced and the exposed film developed and printed. The later addition of light-proof backing paper made the loading and unloading of film an easy task for the camera-owner, and thus the last inconvenience was overcome.

The museum's display of Kodak cameras made by Eastman's company is comprehensive and begins with several of the first production series of box cameras. These are followed by folding pocket cameras with 'extras' such as the autograph device, which allowed the user to title his negatives at the time of exposure, and coupled range-finders for correct focusing.

Standing between the showcases is a huge process camera used for rephotographing photos or drawings preparatory to making zinc blocks in order to print illustrations in newspapers, books, and so on. It has the capacity for enlarging any illustration up to 66 cm across.

In about 1930 the design of cameras showed a sudden change in basic construction. Wood gave way to metal, the size of the negative was greatly reduced, and film winding and shutter operation was greatly simplified. Many of these early-modern cameras are displayed together with the more bizarre evolutions, such as the spy camera. The museum's spy camera is disc-shaped with a diameter of about 13 cm and a small lens near the centre. It was intended for suspension from the neck with the lens pushing through the buttonhole of a jacket.

A section displaying stereoscopic or 'three-dimensional' cameras and viewers illustrates a once popular drawing-room amusement. The process was invented by Charles Wheatstore in 1832. It involves a twin exposure, with a slight difference in camera angle to give the viewer the same impression of relative distance as is produced by the dual vision of human eyes. This process survives today, mainly as a technical aid for doctors, scientists, and engineeers.

Other displays include massive Magic Lanterns and slide projectors powered by gas or pressure oil and projecting from slides 8 x 8 cm to give pictures of much better quality than the electric 35-mm projectors of today.

The enlargers range from enormous nineteenth-century machines to dainty quarter-plate models used horizontally with kerosene lighting. Enlarging and developing equipment include a red darkroom candle lamp, sensitive chemical scales, early daylight loading tanks, exposure meters, and early flash equipment with flint igniters.

The photograph display is perhaps the most compact and detailed exhibition in the museum. It is a miniature museum in itself.

Recorded Sound Section

The Recorded Sound Section of Motat was established in 1968. Various gramophones and music boxes obtained by the museum were displayed at first in the mini-court next to the Pump House. Some twelve months later the section had expanded to such a degree, through donations of more machines as the result of the display and through restoration of machines previously in storage, that new quarters had to be found. Recorded Sound was then shifted into one half of the cottage which it still shares with the Photographic Section.

At the special opening ceremony of the new section, the Auckland radio personality Merv Smith was invited to record his voice on a wax-cylinder Edison machine. The publicity for this event caused a tremendous traffic block and Smith was forced to park his car so far away that he was barely able to arrive on time. The success of this event was such that the section now stages regular live-sound demonstrations during Live Weekends.

A Music Room was later included in the permanent cottage display. This is a working exhibit room from which regularly can be heard a variety of instruments including cylinder and disc-type gramophones, cylinder and disc-type music boxes, and handwind, paper roll and barrel type organs. A 1901 piano-player of (a forerunner of the player piano) with a paper roll mechanism has 65 'fingers' extending from the back which play the keys of an ordinary piano. It is pedal-operated.

The major recorded sound items on display at the museum are:

1890 12-tune cylinder music box with bells, castanets and drum.

1907 39-cm table model Polyphon disc music box.

1907 Cabinetto paper roll organette.

1907 Celestina paper roll organette.

1907 Bacigalupo (German) 8-tune barrel organ.

1907 Aeolian Grande player organ.

1905 Model B Edison Banner Standard phonograph.

1905 Model B Edison Standard phonograph with conversion to play two- and four-minute records.

1908 Model C Edison Standard phonograph with cygnet horn and diamond B reproducer.

1908 Model A Edison fireside phonograph with cygnet horn and diamond B reproducer.

1908 Edison Maroon Gem phonograph.

1913 Edison opera phonograph.

1926 Edison Amberola 30 phonograph.

1929 Edison Oak Chippendale diamond disc phonograph.

1929 Edison belt-drive London Model diamond disc phonograph.

1906 H.M.V. Sheraton extending mahogany horn gramophone.

1908 H.M.V. Monarch Junior extending mahogany horn gramophone.

1925 H.M.V. re-entrant horn cabinet model gramophone.

1925 Lockwood's Perophone external horn phonograph.

1925 Brunswick Panatrope (first all-electric gramophone).

1925 Wurlitzer 24-selection 78 juke box.

1925 Peter-Pan portable gramophone.

1908 Klingsor gramophone.

1901 Aeolian Pianola piano player.

Early office dictaphones, wire-recorders, and magnetic-disc recorders are also on display. Selected items from the extensive collection of Mr L. Stenersen are often displayed.

Motat collection

Motat collection

1. Visitors from the Hokitika area of the South Island, members of the very popular Kokatahi Band entertain in the Transport Pavilion. They play a variety of very unusual instruments.

Music

Music played a great part in the austere life of New Zealand's pioneering families. Social life invariably included sing-songs around the piano, and English traditions of military brass bands and Scottish pipes and drums were inherited from the first immigrants and soldier settlers. Merry-go-round organs provided a joyful atmosphere at shows and fairgrounds, and it is this latter atmosphere that Motat has striven to preserve.

At the museum music brings joy to the 200,000 annual visitors. Before 1970 it was mainly confined to displays of recorded sound instruments and the occasional brass or pipe band played during Live Weekends. However, the hand-turned Celestina Victorian organ, played by a person dressed as an organ-grinder complete with a life-sized cane monkey, was so popular that it became a regular feature on holidays and Live Weekends. And in 1971 a Dutch Souvenir Charity Organ was brought to the museum from the South Island. This large street organ played for three months in Auckland by courtesy of Mr Cornelius de Rijk of Bluff, and it created such a happy carnival atmosphere in the museum grounds that when it returned south it was very sadly missed.

For almost a year music was provided during weekends and holidays by afternoon concerts on an electronic organ, played by the late Mr Lionel Bennett. Notable musical interludes were visits by various musical groups. The Puhoi Bohemian Band, made up of a group of descendants of settlers to the Puhoi district, north of Auckland, who immigrated from Bohemia in the 1860s, play a combination of instruments rarely seen outside Europe. This includes a 'doodlesac', a European form of bagpipes. The Kokatahi Band, which was formed in 1910 by farmers and miners from the Kokatahi Valley, a rich farming area 12 miles east of Hokitika on the West Coast of the South Island, have also played at Motat. Their uniform is based on the 'going out' dress of the early miners — white trousers and brilliant red shirts, with a black neckerchief. Lively old tunes that will never die are played on a variety of unusual instruments. The Mandolinata Orchestra, a string orchestra of mainly mandolins, and a Maori concert party have also performed at Motat. On occasions Country and Western dancing groups and balladeers also entertain visitors.

In 1972 the Dutch Organ returned on lease for a year and became firmly established at the museum. It was a great favourite among visitors, and not a few groups have danced to its music, including the visiting Vienna Boys' Choir. In 1973 the organ was offered to the museum but, in spite of strenous efforts, insufficient finance was available to purchase the instrument. It was a sad day when the Dutch Organ was finally returned to the South Island and much sadder when news came through that the organ had been completely destroyed by fire. The organ left fond memories at Western Springs.

Meantime the Museum Chairman, Captain John Malcolm, was in Europe negotiating the purchase of a fine German street organ from Mr Carl Baum of Hamburg. The organ was shipped to Auckland and permanently established at the museum with the generous assistance of an interest-free loan from Broadlands Finance Company, which is being repaid over a ten-year period from money donated to the organ fund.

Built in 1908, the Wellerhaus 58-tonne organ plays on 33 ruth notes, while three figurines on the front move in time to the music. A repertoire of 34 rolls of music ranges from 'Roll out the Barrel' to classical music such as 'The William Tell Overture'. Fresh rolls of modern music are made for the museum in Germany to update the programmes. When the organ plays, people appear as if it were the Pied Piper. On holidays it plays practically all day long, and a measure of its popularity was evidenced by a recent clearance of the donation box which revealed a substantial cash donation with the following note pinned to it: 'The joy of listening to this beautiful organ brought tears to my eyes'.

A portable barrel organ was also purchased from Mr Baum, and this instrument is very popular at Motat and at fairs and other functions elsewhere in the city.

The Recorded Sound Section has continuous background tapes recorded from the actual vintage instruments on display, and atmosphere in the Transport Pavilion is provided by background effects giving typical sounds of each of the forms of transport displayed there. On holidays and special days a choral group sings sacred music in "The Chapel of the Good Shepherd", and on Live Weekends the village school children love to sing. Motat's vintage radio station (c.1921) broadcasts musical programmes of its era through ground lines to various points in the museum. Music has indeed become very much a part of the museum's presentation.

1. Motat's handsome German Wellerhaus organ of 1902 vintage can be heard at regular intervals every day of the week and contributes greatly to the museum's popularity.

1 Motat collection

Motat collection

NUMISMATICS

The Numismatic Section of Motat deals with the whole spectrum of the manufacture and use of coins, medals, and seals together with many other die-struck items. It is housed in the mini-court, an annex to the pump-house.

A special feature of the display is the unique and particularly interesting selection of punches and dies made by Anton Teutenberg, a sculptor and engraver who made most of the medals and seals used in Auckland from 1866 to the end of World War I. He also carved the gargoyles on the Supreme Court Building and the now demolished Shortland Street Post Office. A particularly fierce gargoyle from the latter is on display at Motat. There are also many of his working drawings and tools, including a portable furnace, which make, in all, a complete nineteenth-century engraver's workshop.

Also on display is Teutenberg's diary of his voyage from Germany to New Zealand. A translation of the diary is being published in serial form in Mintmark, the journal of the Numismatic Society of Auckland.

Among the dies are those engraved by Teutenberg about 1887 for the United Fire Brigades' Association's Five-Year Service Medal, which is still issued in the same design. Also shown is a medal issued by the Pacific Commemorative Society in 1974 to commemorate the centenary of the Auckland Metropolitan Fire Board, the obverse die for which was sunk from Teutenberg's century-old punch.

A series of photographs gives a glimpse of the processes involved in producing modern coins in the Royal Mint. There is also a model of a medieval mint.

A non-numismatic section of closely allied items features a collection of silver trowels, keys, and other souvenirs used by three of Auckland's mayors, Sir James Parr, Mr F. L. Prime, and Mr A. E. T. Devore, in laying foundation stones and opening many Auckland buildings, including the pump-house where the display now rests.

Among the seals are impressions of the Great Seal of New Zealand, the first public seal of the nation, and the seals of the Provinces of New Munster and New Ulster, all three cut by Benjamin Wyon RA, the first in 1841 and the Provincial seals in 1848.

The display is completed by a selection of the trophies awarded by the Numismatic Society of Auckland for research and written papers on numismatics. These are shown together with dies used in striking some of the commemorative medals issued by the Society and a selection of modern coins and proof sets donated to the Society's reference collection, mostly by Mr. G. R. White of Papatoetoe.

1. Art students using the Pioneer Village as their subject.

ADMINISTRATION

On the formation of Motat in July, 1960, a Steering Committee was elected by members, and at a special meeting of members and interested parties the first Board of Management was elected. This Board consisted of ten elected from the members of the museum and ten representing government, local government, and commercial interests. It was responsible for the policy and all-over administration of the museum, the day-today administration being in the hands of an Executive Committee — six members of the Board elected by the Board, plus the Secretary. The Chairman of the Executive Committee fulfilled the duties of the Director, and the committee met regularly once a week or more frequently when required.

Development was difficult. Land at Western Springs, because of rock formation, was not suitable as a site for buildings or rolling stock without considerable and expensive development. Finance was not sufficient to keep pace with the enthusiastic progress of volunteer members, who gave their services in co-ordinating the collections of exhibits given to the museum, in landscaping, in track-laying, and in restoration and administration. A government pound for pound subsidy to the extent of £30,000 had been made available, spread over five years, but this was not repeated and the Executive Committee recommended to the Board that a charge be made for entrance to the museum grounds, an unheard of practice in New Zealand at that time but the only logical means of finance.

The Board accepted this recommendation, and the whole pattern of administration changed. Assistance in the way of administration and supervision was now remunerated, and on a part-time basis the Chairman of the Executive Committee became the first Director. By 1970 the museum had grown to such an extent that it was obvious that a full-time Director was required and that the Rules of the Museum should be altered to be more fitting to an organisation which had become one of Auckland's major attractions.

On the advice of a firm of consultants, a three-tier form of administration was adopted. A Board of Trustees of ten was nominated from prominent business people in New Zealand, to act in perpetuity as Trustees of the assets of the museum and be responsible for the broad policy of the museum and for the appointment of the Director.

The second tier is a Management Committee consisting of three members appointed by the Trustees (one at least being a Trustee), together with six members elected at the annual meeting. This Committee can add to their number three further members, one representing the Armed Forces, one local bodies, and the business community. The balance of the Committee has been excellent. The Trustees have appointed members to the Committee, not necessarily members of the museum, who have business or specialised skills, and the elected members represent a very broad spectrum of museum interests. This Management Committee is responsible for the policy of the museum as laid down by the Trustees and for its day-to-day activities.

The third tier, headed by the Executive Director, is made up of the professional staff, each staff member being a specialist in his own field of endeavour, and the sections of volunteer workers. The Director is responsible to the Management Committee, and the sections, through their chairmen, to the Director.

Regular meetings of Section Chairmen serve as open forums on any matter concerned with the museum. These meetings are chaired by the Chairman of the Management Committee who is the channel of communication with between them and the Trustees.

The system is ideal for coping with a major work force of volunteer members working, under the guidelines of an Incorporated Society, side by side with a small band of paid employees under the control of the Director. It is this system and the enthusiasm with which it is put into practice which have been the main forces in the growth of the museum.

EDUCATION

The first Education Officer to the Museum was appointed by the Auckland Education Board to take up duties in 1973. The number of school parties visiting the museum had increased rapidly and some system was needed to regulate the visits and co-ordinate an educational programme.

Many school groups wished to study one or two topics in depth while others wished to introduce children to the museum by looking over the whole complex. To make this easier, information sheets describing each section of the museum were circulated throughout schools in the Auckland Province under the control of the Auckland Education Board.

MEMBERSHIP OF THE MUSEUM

There are two classes of membership of the Museum Society: a personal membership which, at the time of writing, costs $5.00 per annum and consists of two classifications, members who wish to assist with restoration and give other voluntary assistance to the museum and those who wish to be sustaining members of the Society only; and Corporate membership by business firms and other societies, who pay $25.00 per annum.

Membership of the Society entitles the holder and his immediate family to free admission to Motat and to receive the quarterly *Museum News,* an interesting publication which keeps members abreast of developments.

It is interesting to note that membership is by no means restricted to the immediate vicinity of the museum, but spreads throughout New Zealand and to many overseas countries.

Application for membership of the Society should be made direct to: The Director, MOTAT — The Museum of Transport and Technology of New Zealand (Inc.), Western Springs, Auckland.

LOOKING AHEAD

The future of Motat lies in the past. It is hardly likely that the museum will ever run out of exhibit material, for the technological and scientific advances of our age have created an enormous redundancy of equipment in an ever decreasing span of time. For example, the IBM computer on exhibit at Motat became obsolete in seven short years. It was replaced by a more sophisticated device one sixth of its size and capable of calculating 900 times faster. Throughout human history, the latest developments have always seemed to be the ultimate. But progress is such that the ultimate is never reached.

As a transport museum, Motat faithfully portrays the era from the days of the horse and cart to man's journey through space to the moon. As a museum of technology, however, it has barely scratched the surface. To make a study in depth of the scientific advances even in the course of the last century a large area of exhibition space is required. The museum has most of its exhibition material under cover, but to have everything displayed to the best advantage necessitates further building construction. The large and valuable collection warrants worthy buildings. Finance for this purpose will ever be a problem and seemingly quite without solution unless substantial government assistance can be relied upon.

As Motat is literally bursting at the seams, development must go upward. The rocky substrate which up till now has been a disadvantage will provide sound foundations for large buildings. New buildings will be designed to blend with the old, though the old buildings on the Great North Road frontage will be replaced by a Hall of Science. A multi-storeyed Hall of Technology will house extensive displays, restoration in progress, spacious workshops, cinema, reference library, and administrarive offices. Provision for lecture halls and research will also be made in the future.

The future of Motat is secure and prospects for expansion unlimited.

1

Barry McKay

1. A helicopter view of Motat relating to the Chamberlain Park Golf Course in the foreground, Western Springs sportsground, and Great North Road, with Auckland City, North Head, and Rangitoto in the distance.

INDEX (exhibits)

INDEX (persons)